Hybrid Electric & Alternative Automotive Propulsion

Preface

Welcome to hybrid electric and alternative automotive propulsion.

This book has been written because, as fuel demand and environmental pollution increases, it is important that substitutes are found for traditional methods of vehicle drive. An alternative propulsion vehicle is one that operates using something other than the established petrol or Diesel.

Low carbon technologies are a key component of all automotive qualifications in the UK from Level 1 to Level 4 and finding information to cover the subjects in one title has been difficult until now.

Whether you are a vehicle technician, automotive trainer, student or part of the emergency services, an awareness of current and emerging propulsion sources is vital in order to work on or around these vehicles safely.

Warning, you should not attempt to work on high voltage vehicle systems unless you have had appropriate training. Failure to do so may result in injury or death.

This book embraces new technologies, but also helps cover what can be achieved with traditional propulsion methods. The chapters will introduce health and safety, electrical diagnosis and tooling, followed by technical chapters on the operation of low carbon technologies for alternative propulsion.

It also lays out key terms, points of interest, safety and knowledge checks in order to support the information provided within the text.

Chapters:

Chapter 1 Introduction to Electrical Principles and Safety............**Page 3**

Chapter 2 Hybrid and Electric Vehicles...........................**Page 64**

Chapter 3 Alternative Fuel Vehicles...........................**Page 128**

This book offers:

Ideal support for learners and tutors undertaking automotive qualifications.

Information to help cover the requirements for knowledge of low carbon technologies.

Information to help cover the requirements for the knowledge and understanding of hybrid and electric vehicles.

A large number of illustrations to support knowledge and understanding.

Information that would be useful for emergency services, including police, fire, ambulance and coastguard.

Text © Graham Stoakes 2014

Original illustrations © Graham Stoakes 2014

The rights of Graham Stoakes to be identified as author of this work have been asserted by them in accordance with the Copyright, Designs and Patents Act 1988.

Copyright notice ©

All rights reserved. No part of this publication may be reproduced in any form or by any means (including photocopying or storing it in any medium by electronic means and whether or not transiently or incidentally to some other use of this publication) without the written permission of the copyright owner, except in accordance with the provisions of the Copyright, Designs and Patents Act 1988 or under the terms of a licence issued by the Copyright Licensing Agency, Saffron House, 6 - 10 Kirby Street, London EC1N 8TS (www.cla.co.uk). Applications for the copyright owners written permission should be addressed to the author.

Acknowledgements

Graham Stoakes would like to thank Anita and Holly Stoakes for their support during this project.

Thank you to alerrandre for the cover design.

Author

Graham Stoakes AAE MIMI QTLS is a lecturer and author of college textbooks in automotive engineering for light vehicles and motorcycles.

With his background as a qualified Master Technician, senior automotive manager and specialist diagnostic trainer, he brings 30 years of technical industry experience to this title.

Cover design - fiver.com/alerrandre

Published by Graham Stoakes

First published July 2014

First edition

ISBN 978-0-9929492-0-4

Chapter 1 Introduction to Electrical Principles and Safety

This chapter provides an overview of the health and safety awareness that is required when working on vehicle systems. (Specific health and safety information and control measures relating to high voltage and alternative propulsion systems will be described as appropriate throughout this book). It also introduces the fundamental operating principles of electricity and electrical systems that will aid you when undertaking maintenance and repairs.

Contents

Introduction	4
Legal requirements	5
Fire safety	16
Basic first aid	18
Electrical and electronic principles	22
Electrical units	29
Ohm's law and Power law	32
Types of circuit	34
Electrical and electronic testing	37
Test lamps	37
Power probes	39
Multimeters	43
Inductive clamp meters	48
Oscilloscopes	52
Scan tools and fault code readers	54
Electrical fault finding	57
Preparing for assessment	63

Low Carbon Technologies

Introduction

All work undertaken on vehicles will require general Personal Protective Equipment. The following list can be used to help you decide the type of PPE that you may need, however specific protection will be described at the appropriate points within this book.

Personal Protective Equipment (PPE)

Safety helmet protects the head from bump injuries when working under cars.

Safety goggles reduce the risk of small objects, chemicals or electrical burns to the eyes.

Overalls provide protection from coming into contact with oils and chemicals. Overalls should have non-electrically conductive fasteners.

Safety gloves insulating safety gloves provide protection from high voltage electrical systems used in electric and hybrid drive.

Safety boots protect the feet from a crush injury and often have oil and chemical resistant soles. Safety boots should have a non-metallic protective toe-cap and rubberised soles.

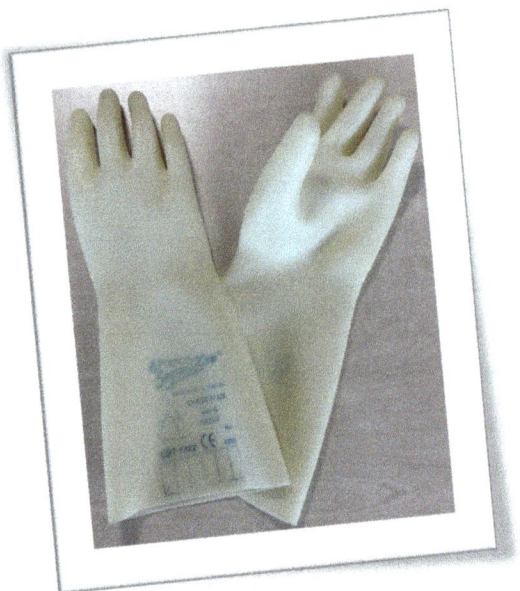

Figure 1.1 High voltage safety gloves

Vehicle Protective Equipment (VPE)

To reduce the possibility of damage to the car, always use the appropriate vehicle protection equipment (VPE):

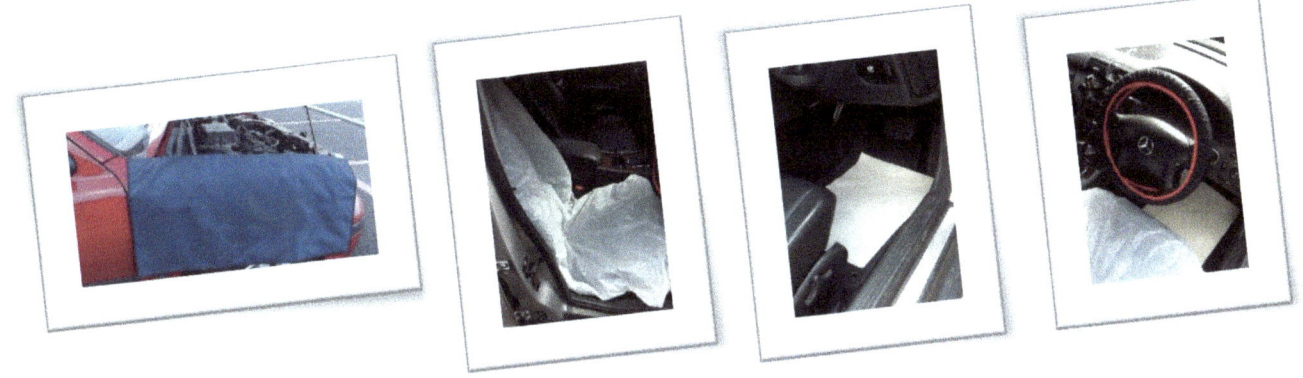

Figure 1.2 Wing covers

Figure 1.3 Seat covers

Figure 1.4 Floor mats

Figure 1.5 Steering wheel covers

Legal requirements

In order to conduct road tests on customer's cars, you must ensure:

- You hold the correct type of driving license for the vehicle being tested
- The vehicle has a valid MOT and tax disc
- You are insured to drive this vehicle on the road
- The vehicle is not clearly in an unroadworthy condition

Health and safety legislation

Health and safety policies are there to help protect you, and will have been developed to make sure that government **legislation** is observed. It is important that you are aware of the legislation and your rights and responsibilities, as well as those of your employer. It is your right to expect your employer to fulfil their responsibilities and it is your employer's right to expect you to fulfil yours. Legislation is the law and, if you do not observe it, you are committing an offence.

Legislation – the laws which have been passed by government and which are enforced by the police and other bodies

The Health and Safety Executive (HSE)

The Health and Safety Executive (HSE) is the national independent watchdog for work-related health, safety and illness. They are an independent regulator and act in the public interest to reduce work-related death and serious injury across all workplaces in the UK. HSE inspectors have powers to issue **improvement notices** and **prohibition notices** if they believe that there are any poor health and safety practices in a workplace they are inspecting.

Improvement notice – notification that the employer must eliminate a risk, for example a bad working practice. The improvement notice gives the employer a specific period of time to eliminate the risk.

Prohibition notice – notification that the employer has to immediately stop all work until the safety risk is eliminated. A prohibition notice is only issued for serious safety risks which involve, or will involve a serious risk of personal injury.

Some of the laws that are relevant to the automotive industry are covered in the next section:

The Health and Safety at Work Act 1974 (HASAWA)

The health and safety of everyone in the workplace is protected by the Health and Safety at Work Act (HASAWA). This law protects you, your employer and all employees while at work. It also protects your customers and the general public when they are visiting your workplace.

Personal Protective Equipment (PPE) at Work Regulations 1992

This regulation requires that employers provide appropriate personal protective clothing and equipment for their employees.

When selecting PPE, make sure that the equipment:

- is the right PPE for the job – ask for advice if you are not sure

- fits correctly – it needs to be adjustable so it fits you properly

- is properly looked after

- prevents or controls the risk for the job you are doing

- does not interfere with the job you are doing

- does not create a new risk, e.g. overheating

- is comfortable enough to wear for the length of time you need it

- does not impair your sight, communication or movement

- is compatible with other PPE worn

The CE mark found on PPE confirms that it has met the safety requirements of the Personal Protective Equipment at Work Regulations 1992. All PPE should have this mark.

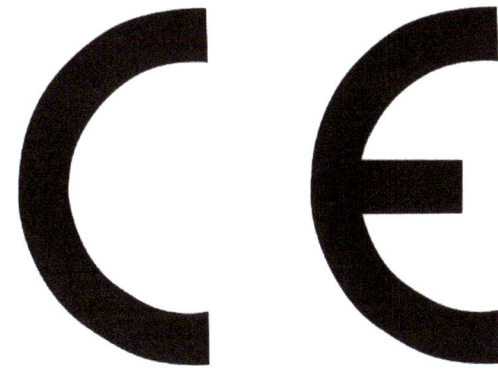

Figure 1.6 The CE mark

Examples of specific PPE related to high voltage or alternative propulsion systems can be found throughout this book.

Although personal protective equipment should be considered mandatory for repairing and maintaining high voltage systems, it is recommended that PPE is used whenever you work on vehicles. However, in many cases, PPE should only be considered a last resort, when risks cannot be adequately controlled or reduced by other methods.

Provision and Use of Work Equipment Regulations 1998 (PUWER)

The equipment used in your workshop needs to be:

• safe to use

• maintained correctly

• inspected regularly

• only used by people who have received appropriate training

The Provision and Use of Work Equipment Regulations 1998 (PUWER) place the responsibility for the safety of workplace equipment on anyone who has control over the use of work equipment, including your employer, you and your colleagues.

On the equipment and machinery that you use at work, you will find a warning label informing you of the dangers it can pose. The labels might be:

• Warnings – make sure you follow the instructions to reduce the risk of damage to the machine and operator.

• Restrictions – follow instructions so that you do not go into areas you are not allowed to go.

• Protective devices – ensure all guards are in place before use.

 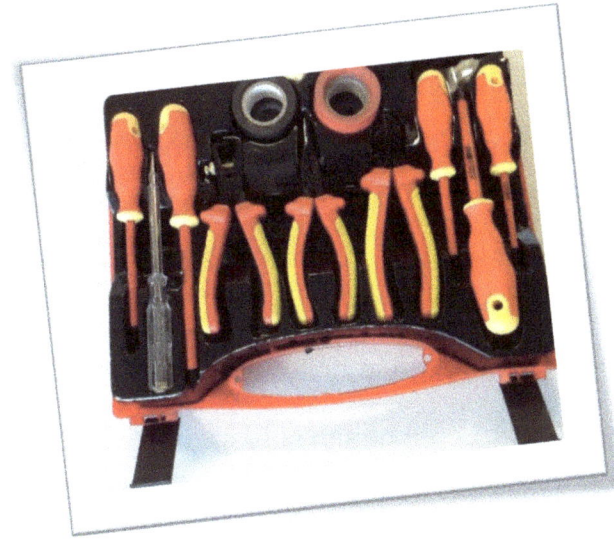

Figure 1.7 Insulated high voltage hand tools

When working on high voltage vehicle electrical systems, remember to use tools and equipment that have been specifically designed for the task. These tools and equipment will normally be highly insulated and, if used correctly, will help reduce the risk of electric shock.

Control of Substances Hazardous to Health Regulations 2002 (COSHH)

The legislation which you and your employer must observe when using hazardous substances in the workshop, is the Control of Substances Hazardous to Health Regulations 2002 (COSHH).

There are eight steps that employers must take to protect employees from hazardous substances. These are shown in the next section:

Step 1: Find out what hazardous substances are used in the workplace and the risks these substances pose to people's health.

Step 2: Decide what precautions are needed before any work starts with hazardous substances.

Step 3: Prevent people being exposed to hazardous substances or, where this is not reasonably practicable, control the exposure.

Step 4: Make sure control measures are used and maintained properly and that safety procedures are followed.

Step 5: If required, monitor exposure of employees to hazardous substances.

Step 6: Carry out health surveillance where assessment has shown that this is necessary or where COSHH makes specific requirements.

Step 7: If required, prepare plans and procedures to deal with accidents, incidents and emergencies.

Step 8: Make sure employees are properly informed, trained and supervised.

Figure 1.8 Hazardous warning labels found on chemicals and dangerous substances

Electricity at Work Regulations 1989

The electricity at work regulations are mainly concerned with electrical appliances, equipment and supply systems, but as the amount of vehicles using high voltages increases, sections of this law may apply to vehicle maintenance and repair. For example, regulation 16 states: 'No person shall be engaged in any work activity where technical knowledge or experience is necessary to prevent danger or, where appropriate, injury, unless he possesses such knowledge or experience, or is under such degree of supervision as may be appropriate having regard to the nature of the work.'

Specific electrical regulations regarding the maintenance and repair of high voltage electrical systems on vehicles can be found in Chapter 2.

Other health and safety regulations

As well as the legislation listed previously, the following regulations apply across the full range of workplaces:

1 The Management of Health and Safety at Work Regulations 1999: require employers to carry out risk assessments, put in place measures to minimise risks, appoint competent people and arrange for appropriate information and training for their staff.

2 Workplace (Health, Safety and Welfare) Regulations 1992: cover a wide range of basic health, safety and welfare issues such as ventilation, heating, lighting, workstations, seating and welfare facilities.

3 Health and Safety (Display Screen Equipment) Regulations 1992: set out requirements for work with computers and visual display units (VDUs).

4 Manual Handling Operations Regulations 1992: cover the moving of objects by hand or bodily force.

5 Health and Safety (First Aid) Regulations 1981: cover the requirements for first aid, including the number of trained first aiders required in the workplace.

6 The Health and Safety Information for Employees Regulations 1989: require employers to display posters telling employees what they need to know about health and safety.

7 Employers' Liability (Compulsory Insurance) Act 1969: require employers to take out insurance against work-related accidents and ill health involving employees and visitors to the premises.

9 Noise at Work Regulations 1989: require employers to take action to protect employees from hearing damage.

10 The Pressure Safety Systems Regulations 2000: users and owners of pressure systems are required to demonstrate that they know the safe operating limits, principally pressure and temperature, of their pressure systems, and that the systems are safe under those conditions. This will include compressed air systems, but does not include pressurised systems used for vehicle propulsion.

- List four pieces of health and safety legislation relating to the automotive industry.
- List three pieces of PPE that should be used when working on a vehicle electrical system.
- Under the electricity at work regulations 1989, what is the purpose of regulation 16?

Accident prevention

Working with high voltage vehicle electrical systems or fuel used in alternative propulsion is extremely hazardous. There are many dangers that could result in an accident causing injury or even death. The management of these **hazards** is key to reducing the risks involved while working on these systems. Not all hazards can be removed, but they can be identified and measures put in place to reduce the dangers that they pose; this is the purpose of a risk assessment.

A risk assessment is an important step in protecting workers and businesses, and is necessary in order to comply with The Management of Health and Safety at Work Regulations 1999. It is designed to focus on the **risks** that really matter in the workplace – the ones with the potential to cause real harm. In many cases, straightforward measures can control risks.

The law does not expect you to eliminate all risk, but you are required to protect people as far as is **reasonably practicable.**

Reasonably practicable – can be carried out without incurring excessive effort or expense.

Hazard – something that has the potential to cause harm or damage.

Risk – the likelihood of the harm or damage actually happening.

A risk assessment is simply a careful examination of what, in your work, could cause harm to people. It allows you to weigh up whether you have taken enough precautions or should do more to prevent harm.

There are five main steps to risk assessment:

Step 1

Identify the hazards – conduct an inspection of your workplace and make a list of all of the hazards you find.

Step 2

Decide who might be harmed and how – remember to include all those who may be at risk, for example:

- staff/colleagues
- contractors
- delivery operatives
- customers
- general public
- young people
- people with disabilities

Step 3

Evaluate the risks and decide on precautions – for example:

- Can you get rid of the hazard completely? If not, what needs to be done to control the risk of harm?
- Is there a less risky option?
- Can the hazard be guarded or access prevented? Is PPE required? (PPE should only be used when other methods of reducing the risk are not practical).

Step 4

Record your findings and implement them – keep a written record of your risk assessment and what you have done to control the hazards.

Step 5

Review your assessment and update if necessary – working situations change, so make sure that you regularly check that your assessment still covers all hazards.

Since the amount of vehicles using high voltage electricity or alternative fuel systems is increasing, you should consider adding these to any current risk assessments.

Environmental protection

Damage to the environment can be caused by contaminating the atmosphere, water supply or drainage system. Under the Environmental Protection Act 1990 (EPA), it is an offence to treat, keep or dispose of **controlled waste** in a way that is likely to pollute the environment (**ecotoxic**) or harm people. People who produce waste must make sure that it is passed only to an authorised person who can transport, recycle or dispose of it safely. You should have procedures in place for working with and disposing of any material which has potential to harm the environment, including high voltage batteries.

Figure 1.9 Symbol for ecotoxic substances

Hybrid Electric & Alternative Automotive Propulsion

Ecotoxic – a substance that is harmful to the environment.

Controlled waste – any waste which cannot be disposed of to landfill, including liquids, asbestos, tyres and waste that has been decontaminated. There are three types of controlled waste listed under the environmental protection Controlled Waste (England and Wales) Regulations 2012: household, industrial and commercial waste. Depending on the classification and type of waste produced, a charge can be made for its collection and disposal.

Environmental Protection (Duty of Care) Regulations 1991

These regulations describe the actions which anyone who produces, imports, keeps, stores, transports, treats, recycles or disposes of controlled waste must take. These people must:

- store the waste safely so that it does not cause pollution or harm anyone.

- transfer it only to someone who is authorised to take it (such as someone who holds a waste management licence or is a registered waste carrier).

- when passing it on to someone else, provide a written description of the waste and fill in a transfer note. (From 2011, the waste transfer note must also include a declaration that you have applied the waste management hierarchy, which means you must consider reusing or recycling your waste before deciding to dispose of it).

- keep these records for two years and provide a copy to the Environment Agency if they ask for one.

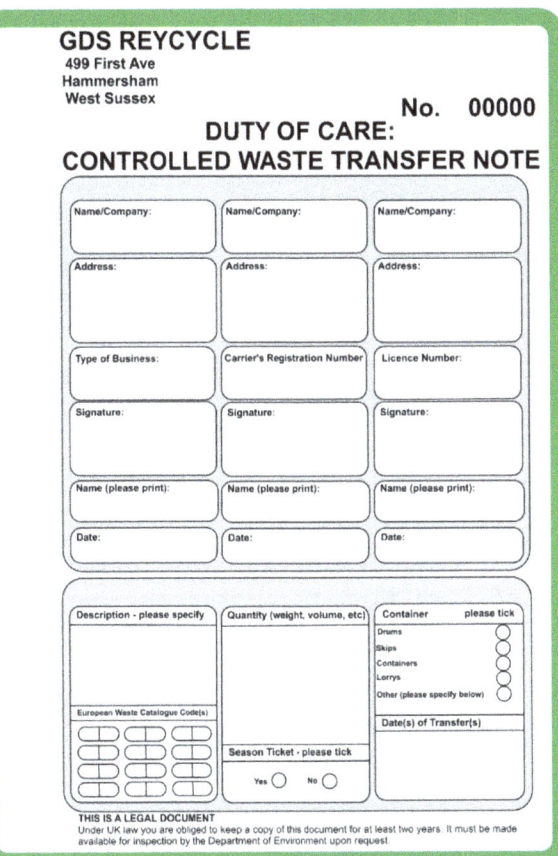

Figure 1.10 Waste transfer note

Environmental protection

During the diagnosis and repair of vehicle electric and hybrid drive systems, you may need to replace batteries. Under the Environmental Protection Act 1990 (EPA), you must treat old batteries as hazardous waste and dispose of them in the correct manner. They should be safely stored in a clearly marked container until they are collected by a licensed recycling company. This company should give you a waste transfer note as the receipt of collection.

Fire safety

Careful consideration should be given to fire safety in the workplace.

- Vehicle high voltage drive systems have an elevated potential to cause fire and provide a source of ignition for flammable materials.
- The substances used as alternative fuels in some vehicles could pose unusual fire risks and may require specialist firefighting knowledge and techniques.

A separate risk assessment should be carried out for fire hazards and safety measures must be put in place.

Figure 1.11 Fire exit sign

Fire extinguishers

In case of fire in the workplace, fire extinguishers should be provided and maintained by the company, for safety. The primary function of a fire extinguisher is to enable you to create an escape route. In any other circumstance you should only attempt to tackle a fire if it is safe to do so and if you have had adequate training.

A number of different fire extinguishers are available depending on the type of fire to be tackled.

The type of fire is normally classified:

Class A: solids, such as paper, wood, plastic etc

Class B: flammable liquids, such as petrol, Diesel, oil etc

Class C: flammable gases such, as propane, butane, methane etc

Class D: metals, such as aluminium, magnesium, titanium etc

Class E: fires involving electrical apparatus

Class F: waste vegetable oil (WVO), and fat etc

Every fire extinguisher has a colour-coded label with a description of its contents and a list of the types of fire it is designed to be used on. Extreme caution should be taken when selecting a suitable extinguisher to use on a vehicle electrical fire. Liquid based fire extinguishers will conduct electricity and this may lead to electrocution.

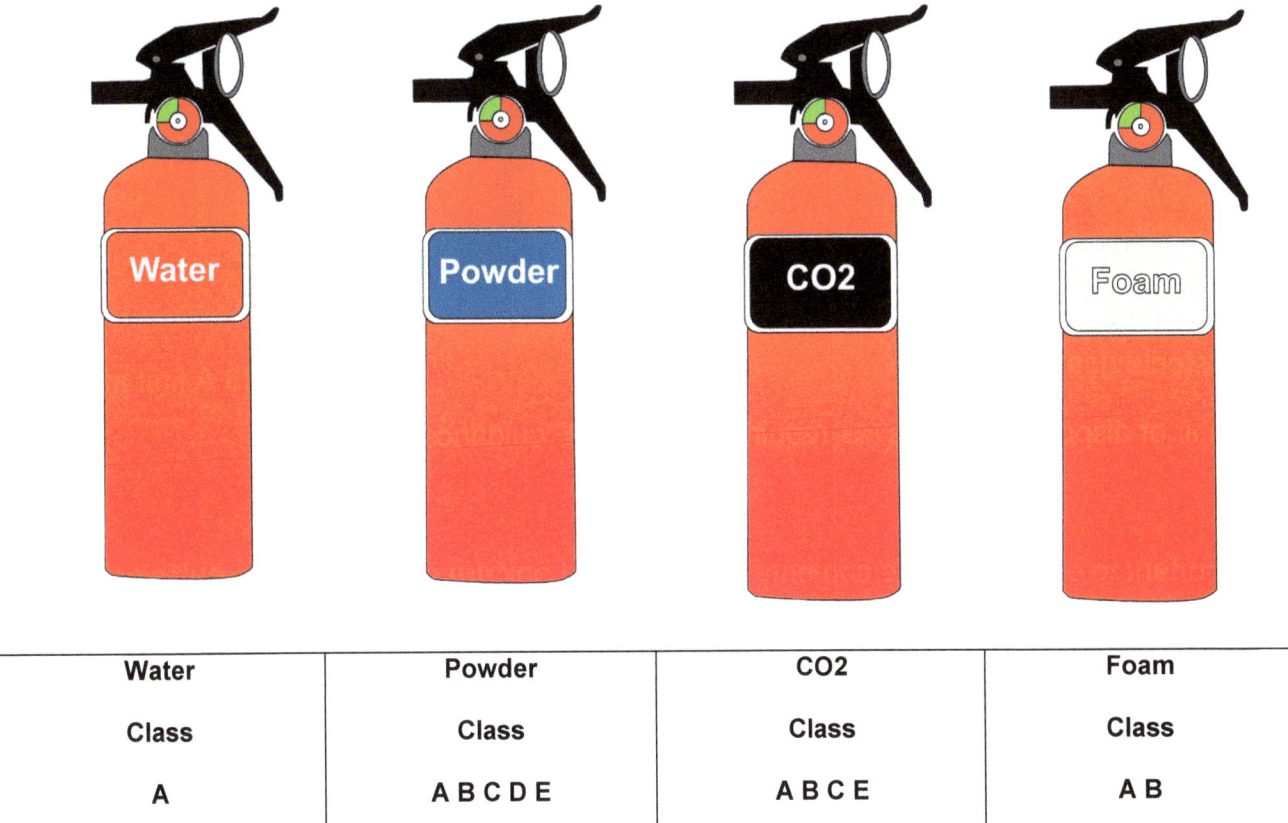

Water	Powder	CO2	Foam
Class	Class	Class	Class
A	A B C D E	A B C E	A B

Figure 1.12 Fire extinguishers

Basic first aid

An automotive workshop is a high risk environment, and no matter what precautions are taken, there is always the possibility of accidents occurring which may lead to personal injury. The following advice is not a substitute for first aid training, and will only give you an overview of the action you may need to take. You should take care when you attempt to administer first aid that you do not place yourself in danger. Be very careful about what you do, because the wrong action can cause more harm to the casualty.

Good first aid always involves summoning appropriate help; many companies will have a trained first aider on site and must have a suitably stocked first aid box.

First aid box

The minimum level of first aid equipment in a suitably stocked first aid box should include:

- a guidance leaflet
- 2 sterile eye pads
- 6 triangular bandages
- 6 safety pins
- 3 extra-large, 2 large and 6 medium-sized sterile unmedicated wound dressings
- 20 sterile adhesive dressings (assorted sizes)
- 1 pair of disposable gloves (as required under HSE guidance)

Figure 1.13 A first aid box

It is important to ensure that the contents of the first aid box are in date and are sufficient, based on the assessment of the workplace's first aid needs.

The law does not state how often the contents of a first aid box should be replaced, but most items, in particular sterile ones, are marked with expiry dates.

Other equipment such as eye wash stations must also be available if the work being carried out requires it.

Getting help

If you need to call for assistance, the main emergency services can be contacted by calling 999 free of charge from any landline or mobile phone.

When calling the emergency services, make sure you give the following information:

- your telephone number
- the location of the incident
- the type of incident
- the gender and age of the casualty
- details of any injuries observed
- any information you have observed about hazards, for example high voltage systems, chemical spills, gas or fuel leaks

The recovery position

When dealing with health emergencies, you may need to place someone in the recovery position. In this position a casualty has the best chance of keeping a clear airway, not inhaling vomit and remaining as safe as possible until help arrives. You should not attempt to put someone in the recovery position if you think they might have back or neck injuries, and it may not be possible if any limbs are fractured.

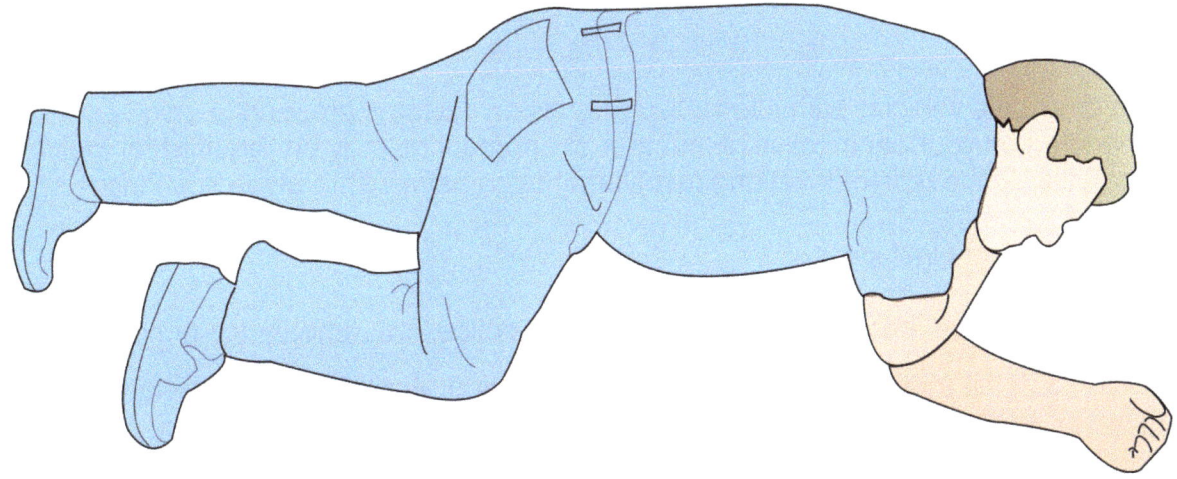

Figure 1.14 The recovery position

Putting a casualty in the recovery position

1. Kneel at one side of the casualty, at about waist level.

2. Tilt the head back – this opens the airway. With the casualty on their back, make sure that their limbs are straight.

3. Bend the casualty's near arm so that it is at right angles to the body. Pull the arm on the far side over the chest and place the back of the hand against the opposite cheek.

4. Use your other hand to roll the casualty towards you by pulling gently on the far leg, just above the knee. This will bring the casualty onto their side.

5. Once the casualty is rolled over, bend the leg at right angles to the body. Make sure the head is tilted well back to keep the airway open.

To find out more about first aid at work, visit the first aid section of the HSE website – http://www.hse.gov.uk/firstaid/index.htm

- ✓ What is the difference between a risk and a hazard?
- ✓ Which two fire extinguishers would be dangerous to use on a class E electrical fire?
- ✓ Describe how you would put a casualty in the recovery position.

Working context

Working on or around vehicles with alternative propulsion systems presents a unique set of issues depending on your perspective. A number of approaches may be required for individual situations relating to the type of level and involvement you have with a given scenario.

Your involvement might include:

- Being a member of the emergency services, such as police, fire, ambulance or coastguard
- Being involved in the motor industry, such as recovery operator, vehicle technician or end of life vehicle recycling
- Being involved in training, such as lecturer, workshop instructor, assessor or workplace mentor

Hybrid Electric & Alternative Automotive Propulsion

The following activity describes a scenario relating to alternative propulsion vehicles and lists a set of suggested precautions or actions that could be undertaken. These actions are not in any particular order or restricted to an individual profession, perspective or approach. Using the information provided in this chapter and the knowledge of your own role or position, create a list of actions that you would take in this situation. Try to put these in a logical order and include any extra steps that you feel are missing or are unique to your role or perspective. (You can also attempt to make lists from the perspectives of other professions or roles from the lists shown on the previous page).

Table 1.1 Working Context

Scenario	Possible steps
• You are the first person on the scene of an accident where someone has been electrocuted while working on a high voltage hybrid electric vehicle. This could be in a workshop or roadside situation.	• Check circulation and begin chest compressions • Isolate the electrics • Summon help • Place victim in the recovery position • Have a suitable fire extinguisher handy • Ensure that the smart key is beyond its range of operation • Wear insulated gloves • Administer first aid • Ensure that others are aware of the situation • Conduct a risk assessment • Remove all sources of ignition • Record injury in the accident book • Report incident to RIDDOR • Allow time for system capacitors to discharge • Tilt the victims head back to open the airway • Listen and check for signs of breathing • Check for burns • Keep the victim warm • Check for any personal medical emergency information • Assess safety of location • Stand on insulating material

When complete, share your list with colleagues or peers to compare your results.

Electrical and electronic principles

The discovery of electricity

Around 2500 years ago a Greek scientist named Thales found that if he rubbed a piece of amber on a piece of cloth, small particles of dust and fluff would be attracted to it. What he had discovered was **static** electricity.

Thales did not understand what was happening, but he did write down his discovery.

The naming of electricity

Around 1550, William Gilbert, Queen Elizabeth 1's doctor, found that if he rubbed a silk cloth on a glass rod it would attract even heavier objects such as feathers. He named this phenomenon electricity after the Greek word for amber, which is elektron.

The first electric current

Unfortunately, static electricity is difficult to turn into a usable source of energy. Electricity needs to move to make it useful. Towards the end of the 18th century, two Italian scientists, Luigi Galvini and Alessandro Volta, created the first moving electricity known as electric **current**. There are two types of current – **alternating current (AC)** and **direct current (DC)**. This current was produced from a chemical reaction and eventually led to the invention of the battery.

Static – stationary, not moving.

Current – moving electricity.

Alternating current (AC) – electricity that moves in two directions (backwards and forwards).

Direct current (DC) – electricity that only moves in one direction.

What is electricity?

Every substance known to man is made of molecules. The molecules of a substance are made up from **atoms**. For example, if the substance is water, the molecule is H2O. This means that the molecule is made up of two hydrogen (H) atoms joined to one oxygen (O) atom.

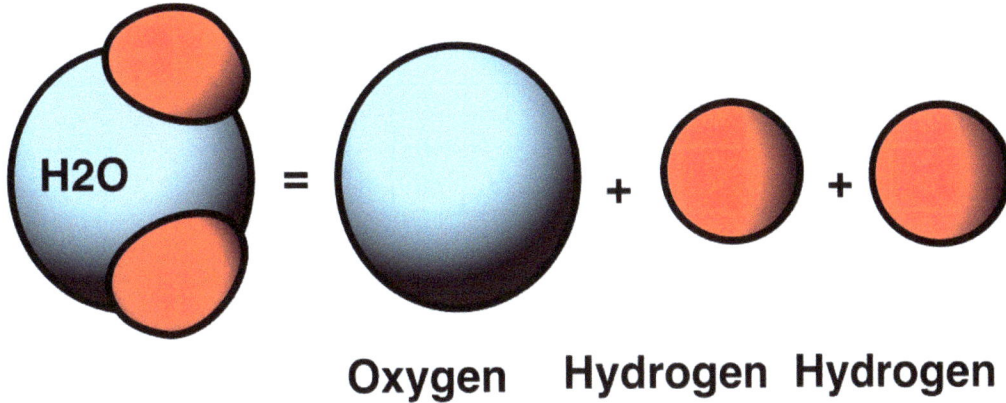

Figure 1.15 The atoms in water

The reason why it can be difficult to understand electricity is because it is contained within atoms. Atoms are very small and hard to imagine.

The easiest way to imagine an atom is like a miniature solar system, with a sun in the middle and planets orbiting around the outside.

In the case of an atom, the nucleus represents the sun. The nucleus is made of positively charged particles known as **protons**. It also contains particles with no charge known as **neutrons**.

Orbiting around this nucleus (in a similar way to the planets) are negatively charged particles known as **electrons**. As the name suggests, it is the electrons that produce electric current.

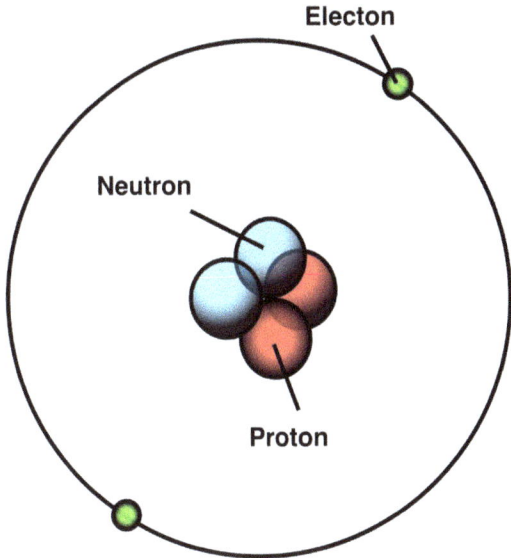

Figure 1.16 A helium atom

Different atoms have different numbers of protons and electrons, as shown in the periodic table.

1 H																	12 He
3 Li	4 Be											5 B	6 C	7 N	8 O	9 F	10 Ne
11 Na	12 Mg											13 Al	14 Si	15 P	16 S	17 Cl	18 Ar
19 K	20 Ca	21 Sc	22 Ti	23 V	24 Cr	25 Mn	26 Fe	27 Co	28 Ni	29 Cu	30 Zn	31 Ga	132 Ge	33 As	34 Se	35 Br	36 Kr
37 Rb	38 Sr	39 Y	40 Zr	41 Nb	42 Mo	43 Tc	44 Ru	45 Rh	46 Pd	47 Ag	48 Cd	49 In	50 Sn	51 Sb	52 Te	53 I	54 Xe
55 Cs	56 Ba	57-71	72 Hf	73 Ta	74 W	75 Re	76 Os	77 Ir	78 Pt	79 Au	80 Hg	81 Tl	82 Pb	83 Bi	84 Po	85 At	86 Rn
87 Fr	88 Ra	89-103	104 Rf	105 Db	106 Sg	107 Bh	108 Hs	109 Mt	110 Ds	111 Rg	112 Cn	113 Uut	114 Fl	115 Uup	116 Lv	117 Uus	118 Uuo

57 La	58 Ce	59 Pr	60 Nd	61 Pm	62 Sm	63 Eu	64 Gd	65 Tb	66 Dy	67 Ho	68 Er	69 Tm	70 Yb	71 Lu
89 Ac	90 Th	91 Pa	92 U	93 Np	94 Pu	95 Am	96 Cm	97 Bk	98 Cf	99 Es	100 Fm	101 Md	102 No	103 Lr

Figure 1.17 The periodic table of elements

Movement of electrons

To make the electric current, you need to move electrons from one atom to the next. To do this they need to be given a push by an external force or pressure.

The pressure used to move electrons can be created by:

- magnets
- a chemical reaction

Orbiting electrons are held in place in a similar way to the gravity acting on the planets circling around the sun. Because the construction of some atoms is so simple, the attraction between the nucleus and the electrons is very strong. This makes it very hard to move electrons. When electrons don't move easily the element is known as an **insulator**.

A copper atom contains 29 electrons and 29 protons. The electrons orbit in circles that get bigger and bigger. The electrons in the farthest orbit have a far weaker bond/attraction to the nucleus than those in a simple atom. These outer electrons are known as 'free electrons'. If an external pressure is applied, electrons can be moved from one atom to the next. This movement of electrons is electric current. When electrons do move easily the element is known as a **conductor**.

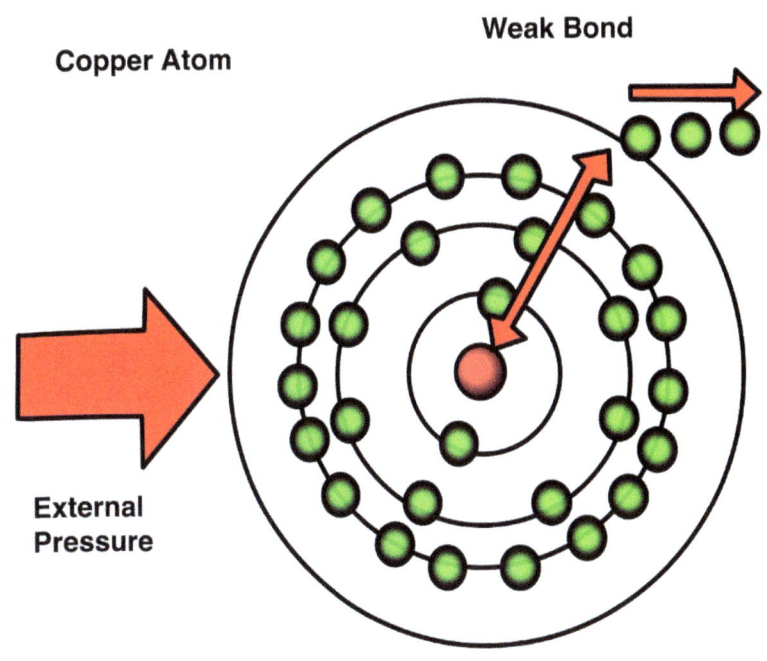

Figure 1.18 A copper atom with external pressure applied

The direction of current flow and electron flow

Electric current will always move from an area of high pressure to an area of low pressure. In conventional or standard electrics, the positive (+) side of the circuit has high pressure, and the negative (−) side of the circuit has low pressure. This means that electricity leaves a battery at the positive terminal, flows through the circuit and then re-enters the battery at the negative terminal. Whenever you test an electric circuit on a vehicle you should use conventional electrics theory (follow the circuit from positive to negative).

True electron flow is in the opposite direction to standard or conventional current (negative to positive). Remember this if the words 'electron flow' are ever used in a description or test.

Low Carbon Technologies

Atom – the smallest component of any chemical element.

Nucleus – the centre of an atom.

Protons – positively charged particles.

Neutrons – particles with no charge.

Electrons – negatively charged particles.

Insulator – a chemical element which does not allow the easy movement of electrons.

Conductor – a chemical element that allows the easy movement of electrons.

Conductors and insulators

• Conductors are used on cars where we want electricity to flow easily, such as wiring.

• Insulators are used on cars to reduce the movement of electricity, such as the coating on the outside of a high voltage cable.

Circuits

For electrons to move from one atom to another, the conductor must be connected in an unbroken loop known as a **circuit**. This means that as one electron leaves it can be replaced by one from behind. If not connected in a circuit, the electrons cannot flow (move), as the last electron in the conductor has nowhere to go. If the circuit is broken it is said to have lost **continuity**.

Figure 1.19 Atoms in a circle to show how electrons will move from one atom to the next

Magnets

Electricity and magnetism are very closely linked. Both electricity and magnets have positive and negative, or North and South, poles. Both **attract** and **repel**.

• If a copper conductor (wire) is passed by a magnet, the magnetic attraction will move electrons through that copper conductor and create electric current.

• If an electric current is passed through a copper conductor then it will generate an invisible magnetic field.

The magnetic effect of electrical current can be used to make things move (by magnetic attraction or repulsion).

That movement can be used to make a motor.

The movement of magnets past a conductor can be used to make electric current. This is the principle of a **generator**.

• Motors turn electrical energy into mechanical energy.

• Generators turn mechanical energy into electrical energy.

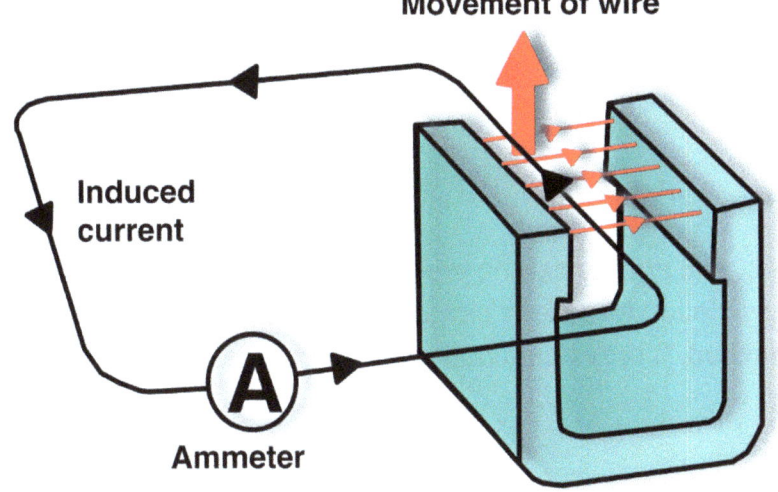

Figure 1.20 The generation of electrical energy by moving a wire through a magnetic field

During charging and discharging of batteries, heat is created. In a high voltage vehicle system, many batteries and charging circuits have their own cooling system to prevent overheating and damage.

Circuit – an unbroken loop.

Continuity – refers to an electrical conductor (something that allows electricity to move easily) which is unbroken or complete (i.e. continuous).

Attraction – bringing together.

Repulsion – pushing away.

Generator – a mechanical component that makes electricity.

Chemical reaction

Electrical energy can be produced by or converted into chemical energy. Because of this it is possible to store electricity and take it with you in the form of a battery. If you keep a battery **charged**, it provides a portable source of electricity that can be used when needed.

The principle of a direct current circuit

As it is very hard to imagine electrons moving from one atom to the other, the process is often best described using water.

Imagine a simple water tower containing:

- a reservoir of water at the top (to represent a battery)
- a pipe leading from the bottom of the reservoir (to represent the wire)
- a tap on the end of the pipe (to represent a switch)
- a small water wheel at the end (to represent a motor)

This analogy can be used to show the operation of a simple electrical circuit. When the tap is turned on, gravity pushes water down through the pipe, under pressure, out through the tap and on to turn the water wheel. This is similar to the way electric current flows through a circuit and turns a motor when the switch is turned on.

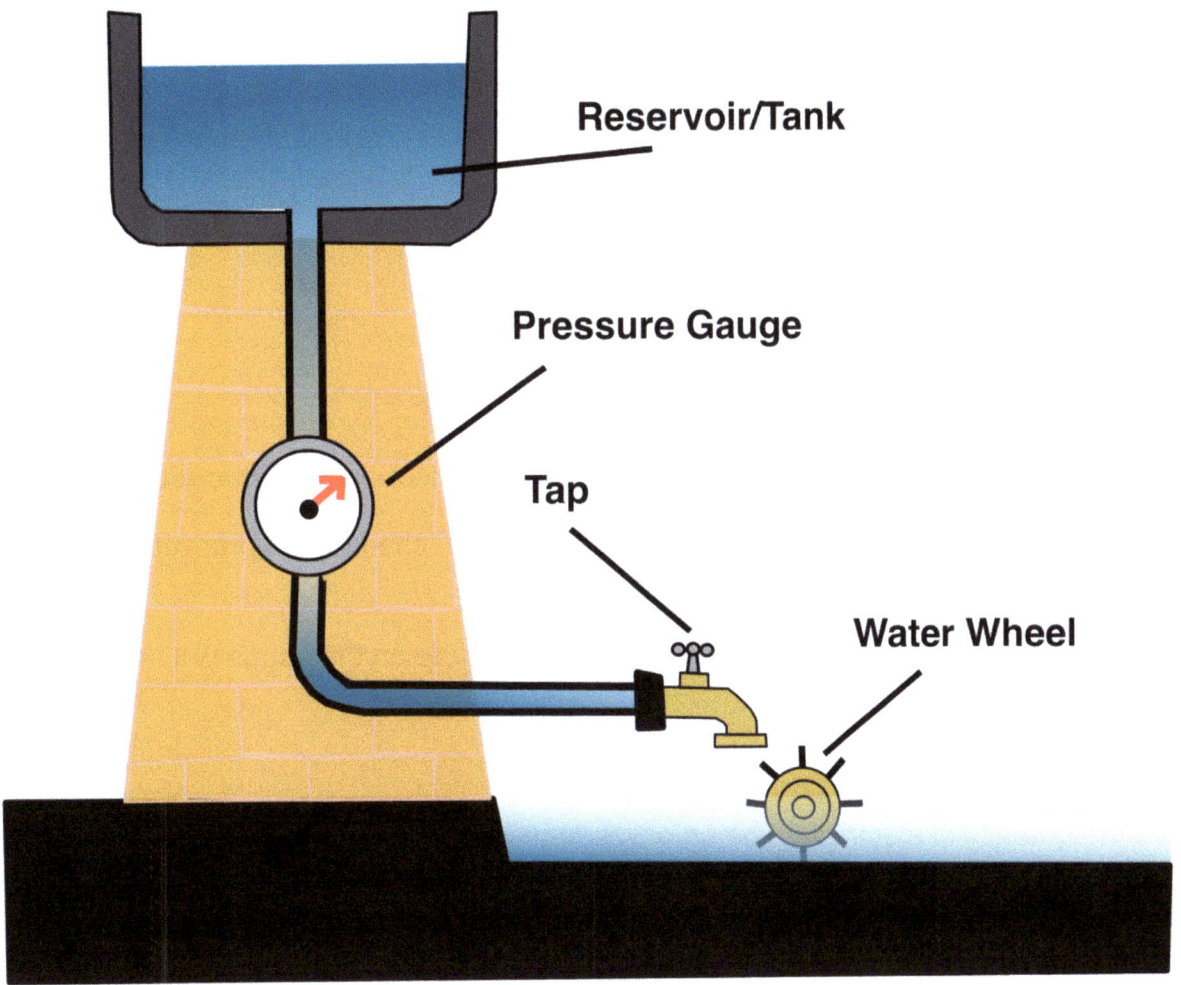

Figure 1.21 A water tower used to represent a direct current (DC) electric circuit

Electrical units

The four main electrical units that you will be using are:

- volts
- amps
- ohms
- watts

They are each named after the person who first described their function.

Volt

Voltage was named after Alessandro Volta. He was the first person to produce moving electricity. Voltage is used to describe the force or pressure in any part of an electrical circuit.

There are two main types of electrical voltage:

- the stored pressure, when everything is switched off.

- the system pressure, when the circuit is switched on.

The stored pressure is known as **EMF** or **electromotive force**.

The pressure found in the circuit when it is switched on is known as **potential difference (Pd)**.

Just as in the water diagram Figure 1.21, when the tap is switched off the pressure gauge will read high (electromotive force).

When the tap is switched on and water can flow, the pressure on the gauge will fall slightly (potential difference).

Amp

Amps are the quantity of electricity. The amp was named after a Frenchman, André-Marie Ampère. Amps are used to describe the amount of electric current (moving electricity) in any part of an electrical system. If you compare them to the water diagram in Figure 1.21, for example, amps are the amount of water flowing down the pipe and coming out the end.

If you think of the water tank as the system's battery, its capacity, or the amount of electricity that it can hold, is measured in **amp hours (Ah)**. Normally, the larger the battery, the more electricity it can hold.

Ohm

An ohm is the **resistance** to electrical flow. Ohms are named after a German mathematician called Georg Ohm. Ohms can be compared to the water diagram in Figure 1.21. The tap at the end of the pipe can increase or reduce the flow of water if you change how far it is opened. If you open it a long way, a large amount of water can flow and this turns the water wheel fast. As the tap is slowly closed, the amount of water coming out of the end of the pipe is reduced. The water wheel turns slower until eventually it stops altogether. The tap is used to control the amount of water flowing in this circuit and it is similar to a 'dimmer switch' found on lighting circuits (a tap for electricity).

Watt

The watt is a measurement of electrical **power** made or used. It is named after a Scottish engineer called James Watt. Compare watts to the water diagram in Figure 1.21. As the water wheel is turned, the energy given up by the water flowing in the circuit to turn the water wheel is measured in watts. If you could turn the water wheel mechanically in the opposite direction, it would be able to scoop up water, forcing the energy backwards and generating current (this would be like the charging circuit). This generation of energy could also be measured in watts.

Volts – unit of electrical pressure.

Electromotive force (EMF) – measurement of electrical voltage with nothing switched on.

Potential difference (Pd) – the difference in electrical voltage when a circuit is switched on and energy is being used.

Amp – unit of electric current which describes quantity.

Amp hours (Ah) – the units used to measure the capacity of the battery.

Ohm – unit of electrical resistance.

Resistance – something that slows down movement.

Watt – unit of electrical power.

Ohm's law

If any one of the units within a circuit (volts, amps, ohms or watts) is changed (i.e. increased or decreased), this will affect all the other units. Using the water tower analogy:

• If the voltage (or pressure) in the water system was increased, more water would flow and the amperage (or quantity) would also increase.

• If the resistance to flow was increased (if the tap was partially closed, for example) then less water would flow and the amperage (or quantity) would also decrease.

This was explained by Georg Ohm with the following mathematical calculations:

amps = volts ÷ resistance

resistance = volts ÷ amps

volts = amps × resistance

With Ohm's law, if you know two of the electrical measurements, you can calculate the third.

The Ohm's law triangle is a good method for calculating the missing unit. It is laid out as shown in Figure 1.22.

In Figure 1.22:

• V = volts (this is sometimes shown as the letter 'E' to represent EMF, but still means volts)

• I = amps (the letter 'I' is used to represent instantaneous current flow)

• R = ohms (the letter 'R' is used for resistance because an 'O' could be confused for a zero).

Figure 1.22 The Ohms Law triangle

How to use the triangle

Cover up the unknown unit with your thumb and you are left with the calculation required. For example, amperage is unknown, so cover the 'I' and you are left with V ÷ R (i.e. volts divided by resistance).

The power triangle

Watts or power can be calculated in a similar way as:

amps = watts ÷ volts

volts = watts ÷ amps

watts = amps × volts

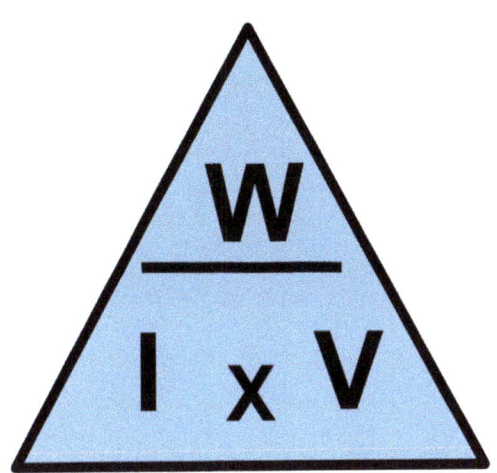

Figure 1.23 The Power Law triangle

A power triangle can be used in the same way as Ohm's law. It is laid out as shown in Figure 1.23

In Figure 1.23:

- W = power (in watts – this is sometimes shown as the letter 'P' to represent power, but still means watts)

- V = volts (this is sometimes shown as the letter 'E' to represent EMF, but still means volts)

- I = amps (the letter 'I' is used to represent instantaneous current flow)

How to use the triangle

Cover up the unknown unit with your thumb, and you are left with the calculation required. For example, amperage is unknown, so cover the 'I' and you are left with W ÷ V (i.e. watts divided by volts).

Low Carbon Technologies

Types of circuit – series and parallel

Series circuits

A series circuit is one where the **consumers** are connected in a line one after another. Where they all follow each other in the same circuit, they share the electricity. Because of this, each consumer will only get part of the voltage available.

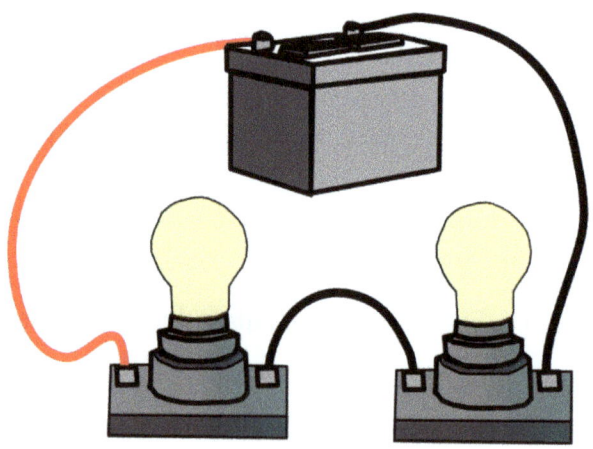

Figure 1.24 A simple series circuit

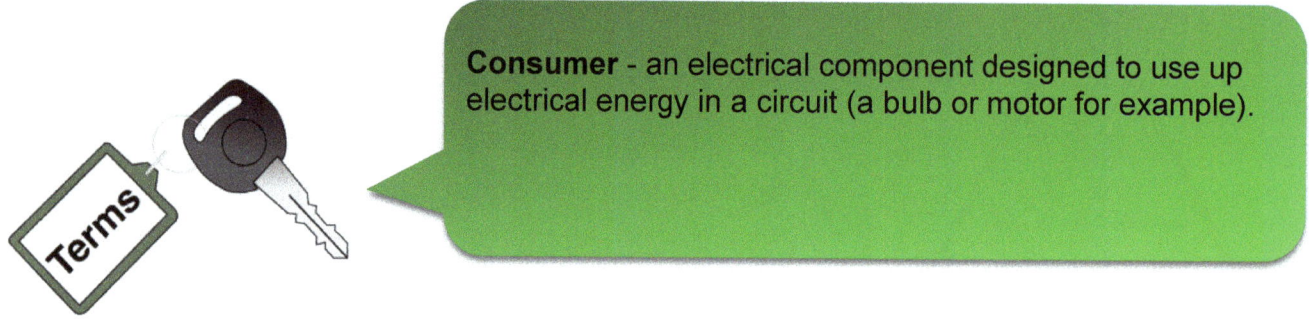

Consumer - an electrical component designed to use up electrical energy in a circuit (a bulb or motor for example).

In a series circuit, if consumers are added, total circuit resistance rises.

In a series circuit there is only one path from the power source through all of the components and back to the source. All the consumers share the electricity, so if more than one consumer is fitted it will only get part of the voltage. Voltage drop will occur across each consumer in the circuit, until all available voltage has been used up. This means that if you connect a voltmeter to a series circuit, you will see the reading on the display fall lower after each consumer. This will continue until you see 0 volts after the last component.

If any one of the consumers fail, the circuit is broken and no electricity can flow. The rest of the consumers stop working. This makes series circuits unsuitable for many systems on cars. For example, if you wired a lighting circuit in series, not only would the bulbs glow dimly, but if one bulb broke all of the others would go out.

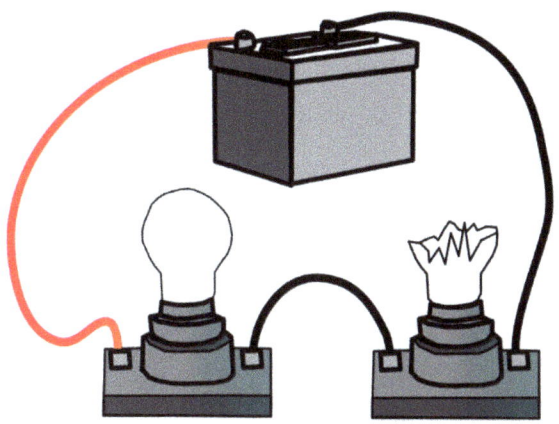

Figure 1.25 When a bulb in a series circuit is damaged the others will no longer work

With a series system, every time a consumer is added, current draw on the power supply will fall.

Parallel circuits

A parallel circuit is one where the consumers are connected next to each other. In a parallel circuit there are multiple parallel paths for the electricity to flow through. Each consumer has its own power supply and earth return back to the power source.

In a parallel circuit, if consumers are added, total circuit resistance falls.

Because each consumer has its own power supply and earth, all the consumers receive the full voltage available and work at full power. So if one consumer in a parallel circuit fails, the others keep working. For example, in a headlight circuit each bulb has its own 12 volt supply and earth return to the battery. If one bulb breaks, the other bulbs will keep working.

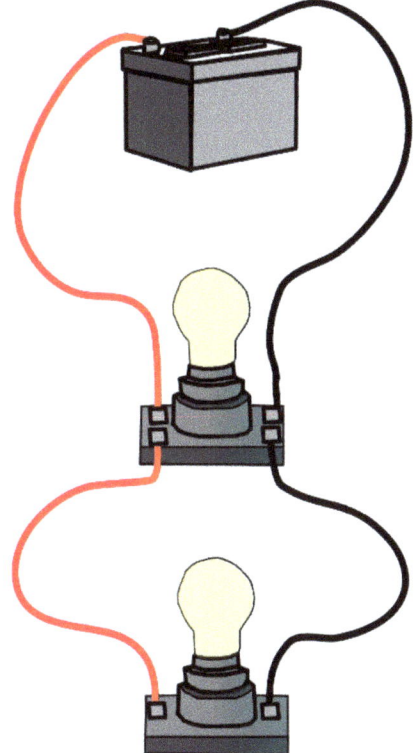

Figure 1.26 A simple parallel circuit

Voltage drop also occurs across the consumers of a parallel circuit. However, unlike in a series circuit, if you connect a voltmeter to a parallel circuit, you will see full supply voltage before each component and 0 volts after it.

With a parallel system, every time a consumer is added, current draw on the power supply will increase.

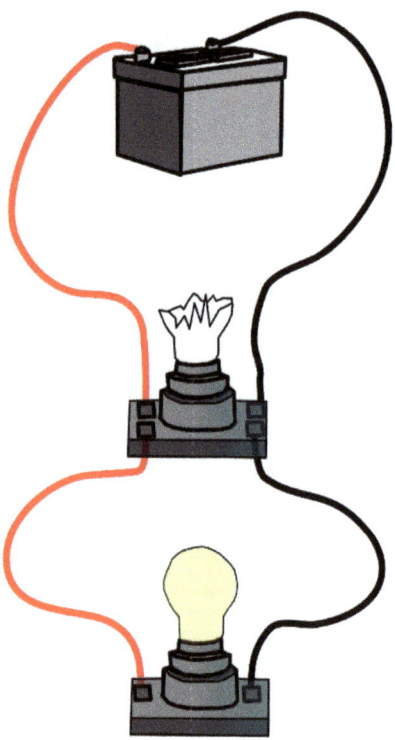

Figure 1.27 When one or more bulbs in a parallel circuit are damaged the others will still work

- ✓ Which component of an atom is used to create electric current?
- ✓ List the four main electrical units.
- ✓ What happens to total resistance of a parallel circuit every time a consumer is added?

Electrical and electronic testing

How to use electrical diagnostic tooling

Electricity is invisible and as a result you will need to use specialist electrical diagnostic tooling to enable you to see what is happening in a circuit. When testing high voltage electrical systems, ensure that the test equipment you are using is suitable, and reduces the risk of electric shock.

Some examples of electrical diagnostic tooling are described in the next section.

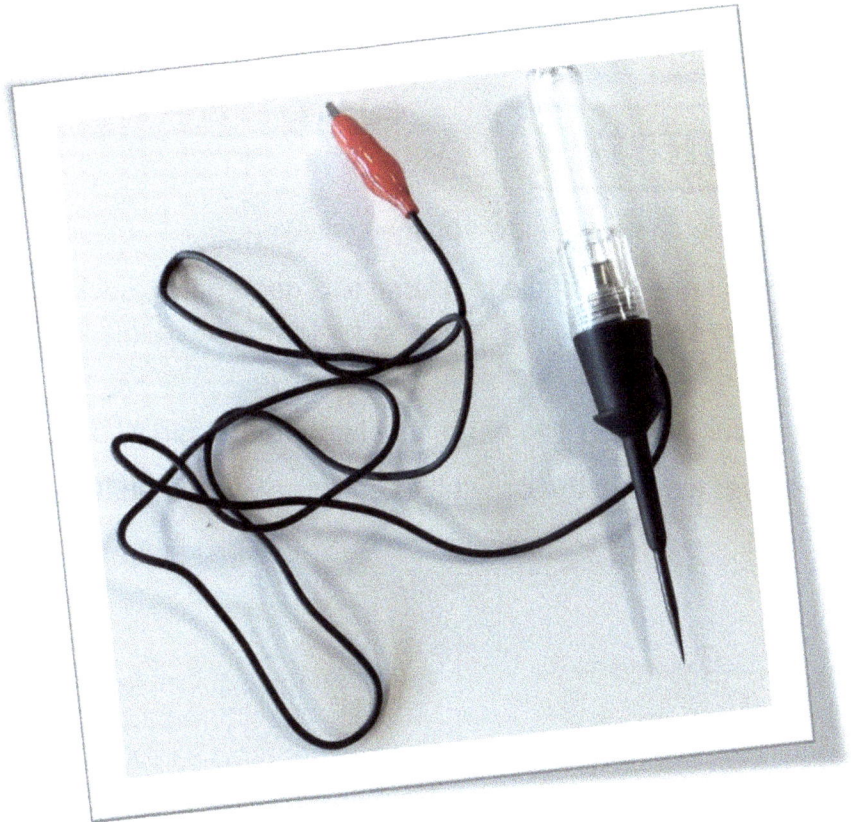

Figure 1.28 A test lamp

Test lamps

One of the simplest diagnostic tools you can use is a test lamp. Whether this is a professionally built tool or a bulb and a couple of pieces of wire that you have put together yourself, this tool can be very effective. Its purpose is to check to see if the circuit has power.

Low Carbon Technologies

A test lamp or LED's are not a suitable diagnostic tools for testing high voltage vehicle systems.

To use a test lamp on a low voltage system:

- Connect one end of the test lamp to a good earth, such as the vehicle chassis or ground (the negative terminal of the battery is better because this is the end of all electrical circuits on a car).
- Connect the other end of the test lamp to the part of circuit that needs to be checked.
- If power exists in the circuit the test lamp will illuminate.

Many modern test lamps have a point at the end of the probe to enable you to pierce wiring installation. Take care when using this probe as it is very easy to stab your finger. Also, if you pierce insulation, you are opening up the wiring to the effects of **oxidisation** from the air. If the wiring is left open, this oxidisation can lead to a high resistance and electrical problems in the future. It is best, where possible, to **back-probe** an electrical plug so as not to damage the wiring.

If you have to pierce the insulation of the wiring, you should cover it with insulation tape or heat shrink. Wiping a small amount of silicon into the pierced insulation hole is not advisable as it can cause corrosion in the wiring and high resistance.

Hybrid Electric & Alternative Automotive Propulsion

Oxidisation – the effect of oxygen on metal, which can cause corrosion.

Back-probe – a method of making a test connection at the back of an electrical socket or plug.

Every time you connect another consumer to an electrical supply wire, more electrical current will be drawn from that supply wire until eventually it can take no more. A test lamp contains a bulb and this is a consumer. A standard test lamp has a low resistance, usually around 6 ohms. This means that when testing a low voltage vehicle electrical circuit, 2 to 3 amps of electrical current may be drawn. If a test lamp is used on an electronic circuit, severe damage can be caused as this high amperage moves through the components.

Always take care when using test lamps to diagnose electrical faults on vehicles. They should only be used when it is safe to do so. It is far safer to use an LED test light, if you are likely to be testing near electronic circuits.

Power probe

A power probe is an advanced form of test light, with additional features and capabilities. Power probes are usually fitted with light emitting diodes, LEDs that are able to illuminate in different colours when connected to either a powered circuit (LED glows red) or an earth circuit (LED glows green).

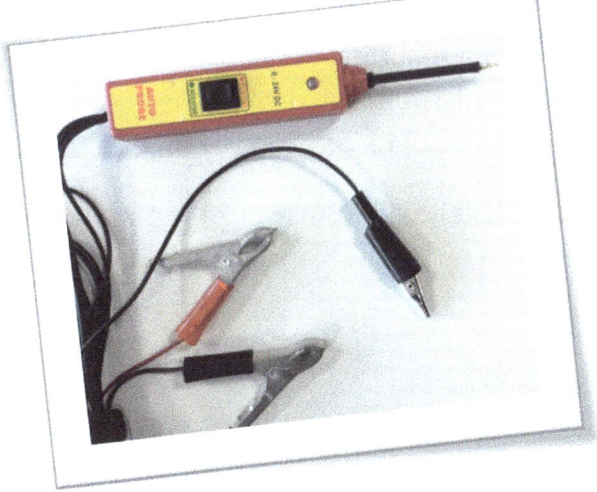

Figure 1.29 A power probe

Low Carbon Technologies

Power probes are not a suitable diagnostic tools for testing high voltage vehicle systems.

Checking polarity

After a simple connection to the vehicle's low voltage battery, you are able to see quite easily whether a circuit is positive or negative, without having to change **polarity** from one battery terminal to another. The power probe normally comes with two crocodile clips (red and black) – connect these to the appropriate positive and negative battery terminals.

Polarity – a term used to describe electrical connection to a circuit. It represents the positive and negative connections.

Auxiliary – something that functions in a supporting capacity.

To check that a correct connection has been made, quickly touch the tip of the power probe to each battery terminal in turn:

- The LED should illuminate red when touched to the positive terminal.
- The LED should illuminate green when touched to the negative terminal.

Checking continuity using a power probe

Not only can a power probe check for electrical feed and earth, you can also use it to check for continuity (a continuous or unbroken conductor). You can check continuity on wires or components that have been disconnected from the vehicle's electrical system.

The power probe has an **auxiliary** ground wire – connect this to one end of the conductor, wire or component. Connect the tip of the power probe to the other end. If continuity exists, the LED on the power probe will illuminate.

Always remember to turn off power first before disconnecting a wire or component on the vehicle's electrical circuit.

Conducting functional tests using a power probe

How to:

You can also use the power probe to undertake functional tests of electrical components.

1. It is recommended that you disconnect the component from the vehicle's electrical system when conducting this test.
2. Connect the auxiliary ground to one terminal of the component and the tip of the power probe should be connected to the other.
3. Check that the LED illuminates to show that the component has continuity.
4. Keeping an eye on the LED, quickly rock the power switch and immediately release.
5. If the LED indicator changed momentarily from green to red, you may proceed with the test.
6. By rocking the power switch forwards and holding it down, electrical potential will be supplied to the component and you can check its operation.
7. If during the initial rocking of the power switch the LED turned off, this normally indicates that the current being drawn by the component is too high for the power probe and the internal circuit breaker has tripped. This may require a manual reset and you will need to check the manufacturer's instructions.

Power probes are only designed to test components which draw relatively small amounts of current. Never use them to test starter motors, vehicle drive motors, etc.

Multimeters

The multimeter is a piece of electrical test equipment designed to measure a number of different units within an electrical circuit. There are two types of multimeter: analogue and digital.

Multimeters can be a suitable diagnostic tool for testing high voltage vehicle systems, but care must be taken when setting up and connecting. Always make sure the the correct scale is selected, that you only hold on to the insulated sections of the test probes and wear appropriate PPE.

Analogue multimeters

Figure 1.30 An analogue multimeter

Analogue multimeters use a needle that moves across a graduated scale to record electrical readings within a circuit. The old-fashioned name for this type of unit was an 'AVO meter', which stood for amps, volts and ohms.

The problem with analogue meters is that they are only as good as the operator. The graduated scale can be difficult to read and so inaccurate readings could be obtained. Depending on the range of the scale provided by the manufacturer, a needle that lies somewhere between two units could be reading any fraction available. Analogue multimeters also have an upper range limit. If the needle flicks all the way to the end of this scale, it is known as full-scale deflection (FSD).

Digital multimeters

Digital multimeters shows digits (numbers) on a liquid crystal display (LCD) screen. These numbers are clearly displayed and are easy to read accurately.

It is quite normal for the last digit on the far right of the screen to continuously change. This is a feature common to most digital multimeters. In a lot of cases, as really high accuracy is not required, this figure can be ignored.

Digital multimeter types

Two types of digital multimeter are common: manually operated and autoranging.

With a manual multimeter, the operator selects the unit and the scale to be measured, normally by turning a dial on the front of the multimeter.

Using a manual multimeter

When you are using a manual multimeter, if you do not know the scale to be used, always follow this procedure:

- For testing volts and amps, first select the highest scale on the dial, then rotate the dial slowly down through the scales until you obtain an accurate reading.
- For testing ohms, first select the lowest scale on the dial and then rotate the dial slowly up until you obtain an accurate reading.

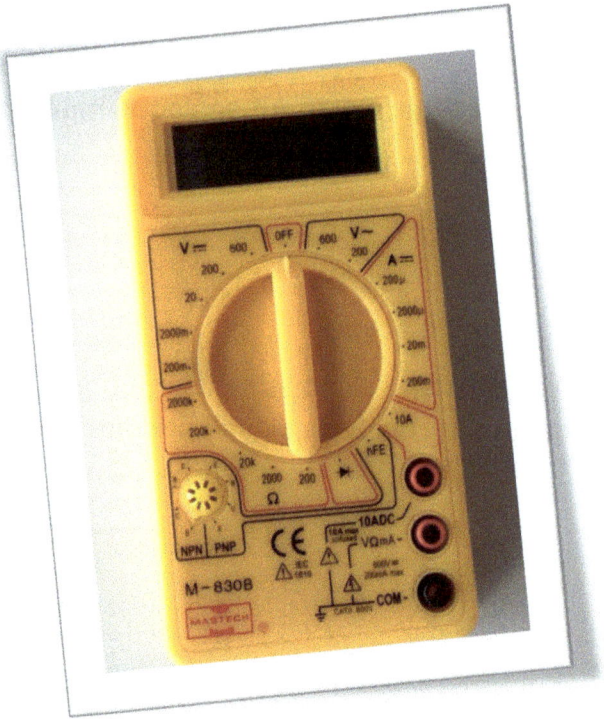

Figure 1.31 A manual digital multimeter (DMM)

Using an autoranging multimeter

With an autoranging multimeter, the operator selects the unit but the scale of that unit is automatically selected by the multimeter. When using an autoranging multimeter, you must take care that your reading is accurate by taking note of the scale of the unit being displayed.

For example, if voltage is measured, the scale might be in:

- Millivolts
- Volts
- Kilovolts
- Megavolts.

Using a digital multimeter

You can measure a number of electrical units on a digital multimeter, including volts, amps and ohms, but other measurements can also be taken.

Extra facilities on a digital multimeter may include: temperature, frequency, diode testing, transistor tests and audio continuity testing.

Figure 1.32 An autoranging digital multimeter (DMM)

The electrical units of volts and amps are often broken down into two further areas: DC === and AC ∼.

- The DC scale is normally shown on the meter as a straight line with a number of dots underneath it ===. This symbol is designed to prevent confusion. If just a single line was used, it might be mistaken for a minus sign and if two lines were used it might be mistaken for an equals sign.

- The AC scale is normally shown on the meter as a wavy line ∼.

- The ohms scale on a multimeter is normally represented by the Greek letter omega (Ω) because if the letter 'O' was used, it might be confused with zero.

Using a multimeter to check voltage

You can use a multimeter as a voltmeter to measure the pressure difference in an electric circuit between where you place the black probe and where you place the red probe.

How to:

1. Connect the probes to the correct sockets on the front of the multimeter.

- Connect the black lead and test probe to the common socket.
- Connect the red probe and test lead to the voltage socket.

2. Most low voltage systems that you will measure on a light vehicle will use direct current DC, so select the scale with the straight and the dotted lines (===).

3. High voltage systems use a combination of alternating current ∼ (AC), for charging and drive systems, and direct current === (DC) at the battery, so ensure that you select the correct type and scale for the circuit you are testing.

4. Following any high voltage warning instructions connect the voltmeter in parallel.

5. Connect the tip of the black lead to a good source of earth, such as the battery terminal, metal bodywork or engine.

6. Use the tip of the red lead to probe the electrical circuit being tested.

Using a multimeter to check for electrical resistance

You can use a multimeter as an ohmmeter to measure resistance. When checking for electrical resistance, always make sure that the power is switched off first, high voltage systems are isolated and any power capacitors have been discharged (see Chapter 2). Disconnect the component to be tested from the circuit.

How to:

1. Connect the probes to the correct sockets on the front of the multimeter.

- Connect the black lead and test probe to the common socket.
- Connect the red probe and test lead to the socket marked with the omega symbol (Ω).

2. Before you take any measurements, you need to calibrate the ohmmeter to check that it is accurate.

3. Turn the selector dial to the lowest ohms setting and join the tips of the two probes together.

4. When the leads are connected, the readout should show zero or very nearly zero. (If any figures are shown on the screen you will need to add or subtract them from your final results).

5. When the leads are disconnected you should see OL (meaning off limits) or the number 1, which is used to represent the letter 'I' (meaning infinity).

6. Now connect the ohmmeter in parallel across the components so that you can measure the resistance.

You can also use the ohmmeter to check for continuity.

To check a piece of wire for continuity - Place the red and black probes at each end of the wire. The screen should display a very low resistance reading.

To check a switch for correct operation - Connect the red and black probes across the terminals and operate the switch. In the off position, the display should read OL (off limits) or infinity. In the on position, the reading on the display should be very close to zero.

Using a multimeter to measure electrical current (Low voltage systems only)

When measuring the electrical current in a circuit use the amps setting on the multimeter, so that it is used as ammeter. Take care when using an ammeter because, if it is connected incorrectly, the multimeter can be damaged.

How to:

1. Connect the probes to the correct sockets on the front of the multimeter.

   - Connect the black lead and test probe to the common socket.
   - Connect the red probe and test lead to the socket used for measuring amps. (This socket is normally separate from the one used to measure volts or ohms).

2. Turn the selector dial to amps measurement.

3. You need to break into the circuit being tested, being careful to avoid short circuits.

4. Connect the ammeter in series, turn on the circuit and measure the current.

A good place to connect an ammeter is at the fuse box – remove the fuse completely and replace it with the ammeter.

Low Carbon Technologies

Never connect an ammeter in parallel (across a circuit). A good ammeter has a very low internal resistance, so if the ammeter is connected in parallel a short circuit is created, causing excessive current flow and the ammeter will be damaged. Also remember that, depending on the quality of your ammeter, the amount of current that you can measure may be restricted to around 10 amps.

Inductive amps measurement

Using an ammeter to check electric current is intrusive and the circuit must be broken. Also, incorrect connection may cause damage to your ammeter. For these reasons, an alternative method of testing for amperage has been developed: some multimeters come with an inductive amps clamp, or this can be purchased separately as an additional unit.

The amps clamp uses **electromagnetic interference (EMI)** to measure current flow within a circuit. It does not require connection in series but is simply clamped around the wire to be tested. When the circuit is switched on and current flows, you can read the amperage measurement from the display. (Make sure that you read the manufacturers operating instructions to know how to connect and read the current clamp).
This is not always as accurate as connecting an ammeter in series, but is quicker and should not cause damage if connected incorrectly. It is also able to take much higher amperage readings than a standard multimeter.

Electromagnetic interference (EMI) – a disturbance that affects an electrical circuit due to either electromagnetic conduction or electromagnetic radiation emitted from an external source. It is also called radio frequency interference (RFI).

> There is normally a plus or minus sign on the amps clamp to show which way round it should be connected to an electric circuit.

Using an inductive clamp meter to measure electrical current (high voltage systems)

Measuring electric current on vehicle high voltage systems presents many dangers. Standard multimeters connected in series will likely be overloaded, leading to damaged equipment, vehicle systems and the possibility of electric shock. During the connection and disconnection of high voltage circuits required to take the amperage readings, there is a high risk of arc flash or arc blast (see Chapter 2, Table 2.5). To reduce the risks involved, it is far safer to use an inductive clamp to measure current, although all recommended safety precautions should still be taken.

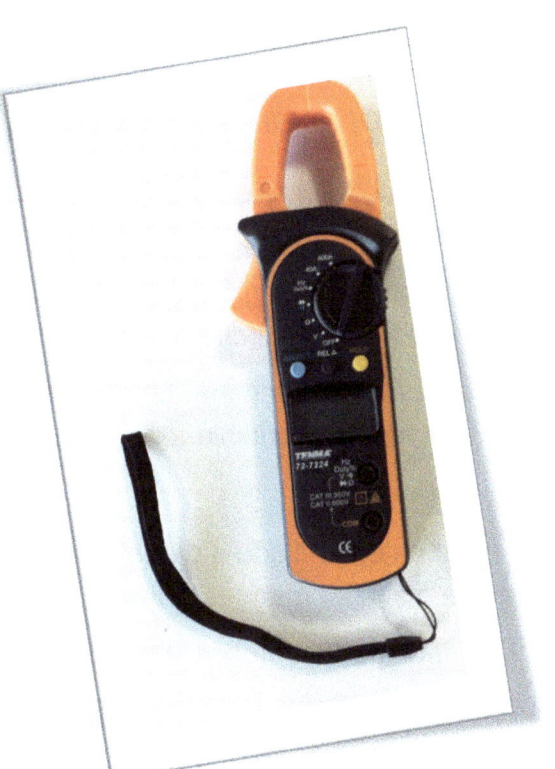

Figure 1.33 An inductive clamp ammeter

How to:

1. Identify the high voltage system and any manufacturer safety precautions.
2. Determine if the system to be tested is operating with alternating current AC $\sim$ or direct current DC $===$.
3. Following manufacturer safety precautions including high voltage PPE, connect the inductive clamp around, the wiring to be tested.
4. Set the scale to the highest reading available and operate the system.
5. Adjust the scale until an accurate reading is obtained and compare with manufacturers recommendations.

Other functions of a multimeter

Many multimeters are capable of other functional tests in addition to checking voltage, amperage and resistance. Some examples of extra functions are described in the next section.

Audible continuity testing

Some multimeters include an audible continuity tester. This means that you can test the continuity of an electrical component without having to look at the screen.

- Connect the test probes to the multimeter: black to the common or ground socket and red to the ohms socket.
- Turn the dial to the audible continuity test setting.
- To calibrate the meter and check correct operation, touch the probes together. You should hear an audible tone.
- As with ohms testing, you must switch off circuit power and remove the component being checked from the circuit.
- Now connect the red and black probes to the terminals of the conductor. If continuity exists, you will hear the audible tone.

Diode testing

Most multimeters include a diode test facility. A diode is a one-way valve for electricity. Conduct the test in a similar manner to the continuity test.

- Connect the test probes: black to the common or ground socket and red to the ohms socket.
- Turn the dial to the diode testing setting.
- To calibrate the meter and check correct operation, touch the probes together. The display should show an ohms reading of zero.
- As with ohms testing, you must switch off circuit power and remove the diode from the circuit. You may need to unsolder the diode to remove it.
- With the diode removed, connect the probes to the terminals. If the diode is operating correctly, the display should show a low ohms reading.
- When the polarity of the probes is swapped over the display should show an off limits or infinity reading.
  - If it shows zero in both directions, the diode has become short circuited.
  - If it shows off limits or infinity in both directions, the diode has become open circuited.

Frequency testing

Some multimeters have a **frequency** test facility. Frequency is a measurement of how quickly a circuit switches. The reading is normally measured in **hertz** (Hz). 1Hz is equal to one complete cycle of operation (on and off for example) occurring in one second.

- Connect the test probe leads to the appropriate sockets on the multimeter.
- Turn the dial to the frequency setting.
- Test the component while the circuit is operating.

Frequency – how often something happens.

Hertz – a measurement of frequency.

Transistor – an electronic component which can operate as a switch or amplifier (with no moving parts).

Temperature measurement

Some multimeters have a temperature measurement facility. This normally requires an additional probe to be connected. The temperature probe usually has its own socket for connection. Once you have turned the dial to the appropriate setting, you can measure temperature by placing the end of the probe where the measurement is to be taken. (Temperature measurement can be useful for diagnosing high voltage cooling system faults for example).

Transistor testing

Some multimeters have a **transistor** testing facility. This facility is rarely used by automotive technicians.

Transistors are small electronic switches with no moving parts. They are normally soldered to an electrical circuit board and have three connections: collector, emitter, and base.

There are two types of transistor in common use: positive negative positive (PNP) and negative positive negative (NPN). If the multimeter has a transistor test facility, a six connector socket will be available, marked PNP or NPN. The transistor must be unsoldered from its circuit and connected to one of these diagnostic sockets. The transistor can now be tested by following the multimeter manufacturer's instructions.

Oscilloscopes

An oscilloscope is a piece of electrical test equipment designed to act like a voltmeter or an ammeter. A multimeter's measurement readout can't change fast enough to deal with modern electronic systems on motor vehicles – the numbers on the screen can't keep up. The answer to this is to use an oscilloscope.

Unlike a voltmeter, oscilloscopes not only show volts or amps but also time. Instead of a digital readout, the results are shown as a graph of volts or amps against time on a screen (as shown in Figure 1.35)

Figure 1.34 Handheld oscilloscope

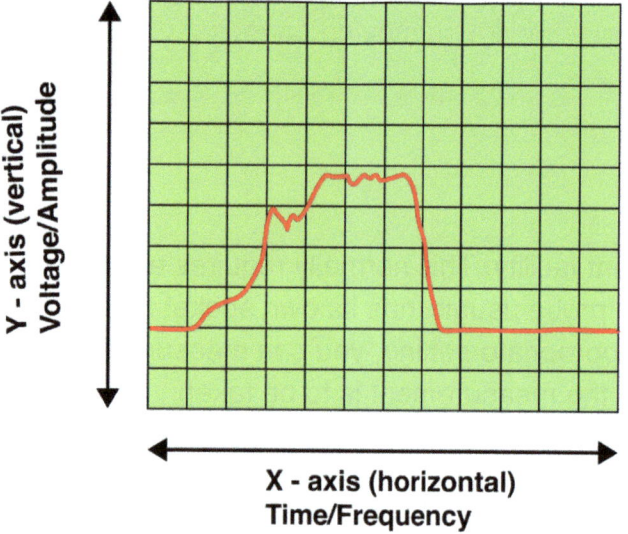

Figure 1.35 An oscilloscope screen

- The graph normally shows voltage or amperage at the side of the screen (on the y-axis) – this axis is often called **amplitude**. Use the scale setting switch in a similar way to the dial on a manual multimeter to choose the amount of volts or amps that are shown on the screen.

- The graph normally shows time across the bottom of the screen (on the x-axis). This axis is often called **frequency**. Use the timescale switch in a similar way to the dial that is used to choose the amount of volts on a multimeter.

An easy way to remember which axis is which on a graph is to say 'X is across' (a cross).

Lots of people are put off using oscilloscopes by the large box containing many wires and connectors. They feel that it will be complicated and time consuming to set up, so they don't bother.

However, to use an oscilloscope for simple electrical testing, you only need two probes – a common and voltage wire – just like a multimeter. To measure amperage, you may need an inductive clamp.

Most of the diagnostic sockets found on oscilloscopes are colour-coded, so after a quick check of the manufacturer's instructions, it should be fairly easy to know where to plug these probes in.

Amplitude – the height of a waveform, measured in volts or amps.

Frequency – the time scale of a waveform (how often something happens).

Using an oscilloscope for electrical testing

How to:

Note: The oscilloscope probes may come in different colours, but for the sake of simplicity we will call them red and black here.

1. Connect the tip of the black lead to a good source of earth, such as the battery terminal, metal bodywork or engine. This will then only leave you with the red wire to worry about.

2. Observing any manufacturers high voltage safety precautions, connect the red probe to the circuit to be tested. (Some oscilloscopes will require the fitting of a resistor known as an attenuator in order to read high voltages).

3. Adjust the scales until you see an image on the screen.

4. After some practice, you will become familiar with the patterns and waveforms created by different vehicle systems.

If you don't know what voltage or timescale to use on an oscilloscope, find out in the same way as you would with a multimeter. Start with the highest setting available and work downwards until you can see an image on the screen.

Scan tools and fault code readers

Faults with many modern vehicle systems would be difficult to diagnose without the aid of a scan tool. The electronic processes that take place within electrical and electronic circuits mean that these systems are being controlled many thousands of times a second, and faults can occur so quickly that you could miss them.

Since the 1980s, manufacturers have been including on-board diagnostic (OBD) systems as part of their vehicle design. The computers that control the vehicle's electrical systems have a self-diagnosis feature. This allows them to detect certain faults and store a code number. Because these electronic control units (ECUs) are monitoring functions, they are able to record intermittent faults and store them in a keep alive memory (KAM) for retrieval by a diagnostic trouble code (DTC) reader.

Hybrid Electric & Alternative Automotive Propulsion

Figure 1.36 Scan tool

It is a common misunderstanding to think that plugging a fault code reader into the vehicle's OBD system will tell you what the fault is. It actually only points you in the direction of the fault. You must test the system and components to find the fault.

E-OBD

European legislation states that any faults with an engine management system which might lead to excessive exhaust pollutants being released to atmosphere must be stored as a diagnostic trouble code. A standardised list of codes and a diagnostic connector were produced to be used by manufacturers selling cars in Europe. In this way, information was made available to all service and maintenance repair facilities. This system has become known as E-OBD.

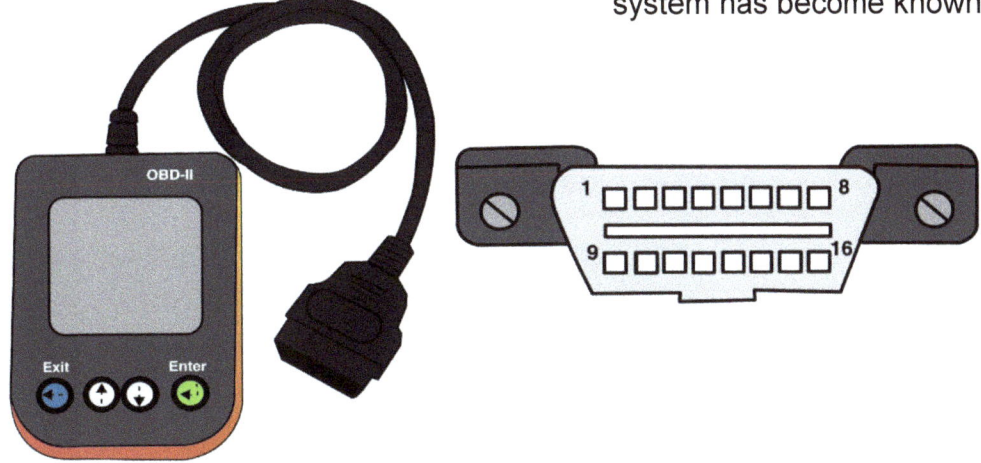

Figure 1.37 A basic E-OBD scan tool and standardised 16-pin diagnostic socket

E-OBD is designed to detect emission-related faults only and should not be confused with original equipment manufacturer (OEM) serial data or diagnostic trouble codes (DTCs). Currently E-OBD data is restricted to the fuels, petrol, diesel and liquefied petroleum gas LPG, although hybrid drive vehicles will comply for their internal combustion engines.

Vehicle types which are E-OBD compliant include:

Petrol engine vehicles:

- vehicles **type approved** from January 2000
- all new car registrations from January 2001

Diesel engine vehicles:

- all new car registrations from January 2003

Liquefied petroleum gas (LPG) vehicles:

- all new car registrations from January 2005

Type approved – refers to a product which has met the legal requirements to be sold in a particular country. Type approval is granted to a product that meets a minimum set of regulatory, technical and safety requirements.

Features of scan tools

Typical features of scan tools include:

- retrieval of electronic control unit (ECU) fault codes
- erasing of system ECU fault codes
- displaying serial data/live data
- displaying readiness monitors
- resetting of ECU adaptions
- displaying freeze frame data
- coding of new components, such as fuel injectors
- access to information on various vehicle electronic systems
- resetting of service reminder lights
- vehicle key coding

Fault finding

Common electrical faults

Diagnosing electrical faults can be confusing, as the symptoms can be very wide and varied. If you follow a simple approach, the diagnosis can be reduced to four main electrical faults:

- open circuit
- high resistance (including bad earth)
- short circuit
- parasitic drain

Open circuit

In an open circuit, electricity cannot flow. This is normally because there is a physical break in the system. As a potential difference Pd will only occur in a circuit when current can flow, the fault can be diagnosed using a test method known as volts drop.

To diagnose an open circuit, you can use the multimeter as a voltmeter. Once set up correctly and connected to the appropriate circuit, measurements can be taken.

If the circuit is working properly, you should see full voltage all the way up to the consumer, at which point the electrical pressure should be used up. In an open circuit, the voltage will disappear. Using a voltmeter you can see at what point in the circuit this happens.

For example, Figure 1.38 shows using a voltmeter to check a low voltage open circuit in which the bulb does not light up. The voltmeter is connected at various points of the circuit to find out the voltage at these points. Where the voltage is different (between 12 volts and 0 volts), this shows the position of the open circuit. In this example, the open circuit is between points B and C.

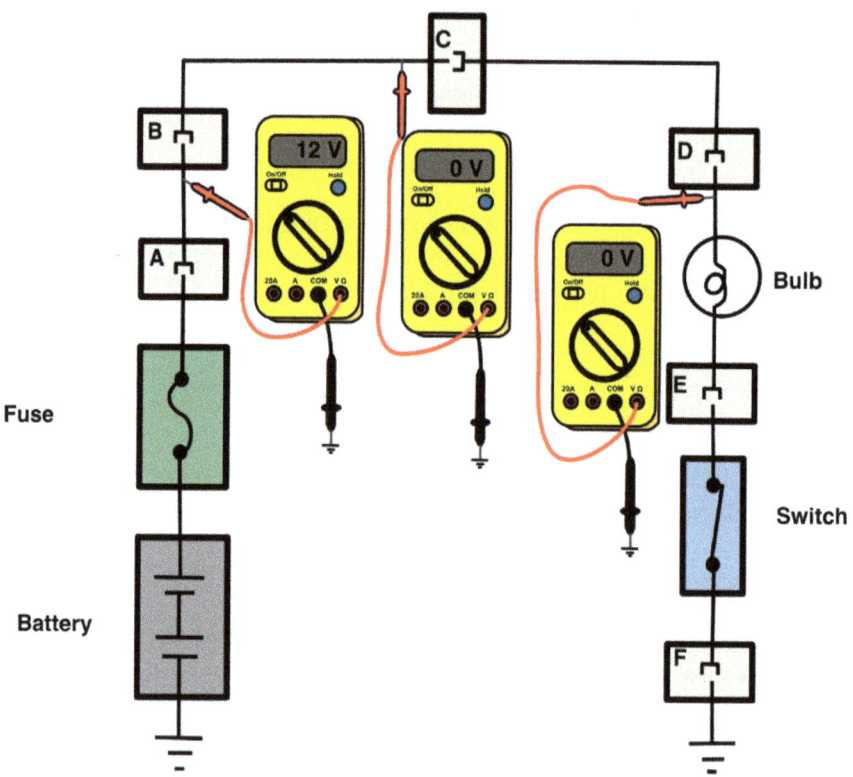

Figure 1.38 Using a voltmeter to check an open electric circuit

High resistance

In a high resistance circuit, the electricity slows down. This is normally because of a partial restriction in the system. Many high resistance faults are caused by poor, corroded or loose connections. A potential difference Pd will occur in a high resistance circuit, but the total Pd will be shared between the consumer and high resistance. This fault can also be diagnosed using the test method known as volts drop.

The symptoms of high resistance are that the component does not work properly (e.g. a bulb that glows dimly) because the circuit pressure (voltage) is shared with the resistance.

To diagnose this fault, you can use the multimeter as a voltmeter. Once set up correctly and connected to the appropriate circuit, measurements can be taken.

If the circuit is working properly, you should see full voltage all the way up to the consumer, at which point all the electrical pressure should be used up. In a high resistance circuit, the full voltage potential difference is not used up by the consumers. Using a voltmeter you can see at what point in the circuit this happens.

If there is a lower than expected voltage at the consumer, then the fault is in the first half of the circuit. If full voltage is present at the consumer, but some voltage still appears after the component, then the fault is in the second half of the circuit (bad earth).

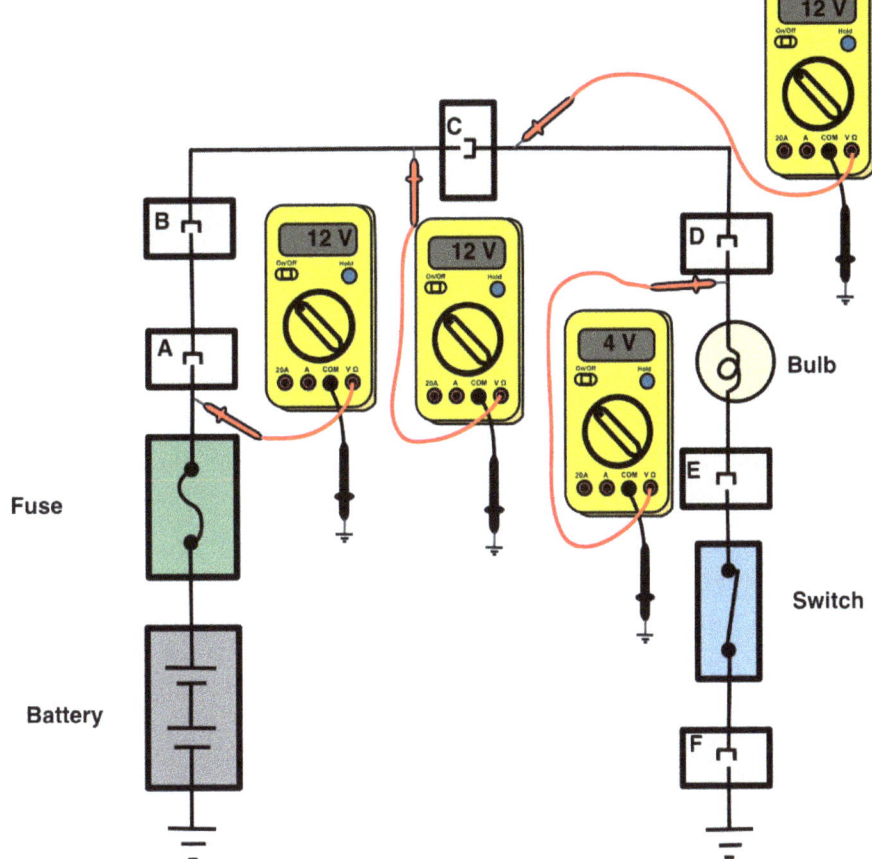

Figure 1.39 Using a voltmeter to check a high resistance electric circuit

Bad earth

A bad earth is a high resistance after the consumer. If this exists the symptoms will be that the component won't work properly. Sometimes a bad earth can also cause the electrical energy to find an alternative path to the negative side of the battery. If this happens you may see symptoms such as all of the lights in the same unit operating at the same time (e.g. brake lights flashing with the indicators).

To diagnose a bad earth use the same procedure as for high resistance.

Short circuit

Electricity is lazy, and will always take the path of least resistance. (Why travel the full length of the circuit when it can take a shortcut?)

In a short circuit, the electricity doesn't make it all the way to the end. Instead of going through the consumer, the electricity makes its way back to the battery early, and in the process converts its energy to heat.

The sudden discharge of current can cause a lot of damage, so the fuse that is used to protect the system should blow. If this happens the symptoms can make you think that the problem is an open circuit (which it is in a way, as the blown fuse has broken the circuit so that no current can flow).

Figure 1.40 An electric circuit with a test lamp bulb connected in the place of the fuse to check for an electrical short circuit

In this situation, you can test the system with a voltmeter as explained in testing an open circuit. But once you have discovered the blown fuse, you should change your diagnostic routine to look for a short circuit. Any heat damage, including blown fuses, is a good indication that a short circuit might exist.

In a low voltage circuit, if a **dead short to earth** exists (e.g. the insulation of a wire has **chafed** against the metal bodywork of the vehicle), you can use a test lamp to help diagnose this fault (see Figure 1.40). It is important to use a test lamp containing a bulb and not an LED, as this could lead to system damage.

Once connected in place of the fuse, if the test lamp illuminates then the electricity is finding an alternative path back to the battery (short circuit). As the bulb is an electrical consumer, it uses up the electrical potential, and shouldn't damage the rest of the circuit. The circuit should then be disconnected systematically from the far end, working back towards the fuse box. When the bulb goes out, you have located the position of the dead short.

When checking for an electrical short circuit, only bypass the fuse with an electrical consumer like a bulb. Using other electrical components (such as wire) could cause a sudden discharge of electricity that may burn you. Remember not to use a test lamp on electronic systems, as the current draw may cause damage.

Checking for a high voltage short circuit

High voltage short circuits are dangerous and pose a very high risk to health and safety. If a short circuit occurs in a high voltage system, you should follow all recommended health and safety procedures, and isolate the power source (allowing any capacitors to discharge so that the system is safe to inspect). A careful visual examination of the system wiring and components should then be conducted. A high voltage discharge will normally show visible signs of heat damage, and if this is found, it should be correctly rectified before the fuse/circuit breaker is replaced and the power reconnected.

Dead short to earth – a direct connection between the positive and negative sides of an electrical circuit.

Chafed – rubbed against something so that the surface has been damaged.

Parasitic drain

A parasitic drain is similar to a short circuit – electricity will continue to flow even if the system is switched off, although this fault may not cause visible system damage. The symptom reported is normally that the battery goes flat if left for a period of time. To help diagnose this fault, you can use the multimeter or an inductive clamp meter as an ammeter. (This acts like a flow gauge to measure the amount of electric current moving in a circuit).

Checking a parasitic drain

To check for parasitic drain, switch off all electric systems and connect the ammeter. The ammeter must be inserted into the circuit (connected in series) so that it isn't damaged. To do this you may need to disconnect one lead from the battery and use the ammeter to bridge the gap, so that current flows through it.

An inductive clamp can be placed around the battery wiring without disconnecting and inserting in series. This is much safer but is often less accurate than using an ammeter directly.

With everything switched off, there should be no current on the display of the meter. If any current (measured in amps) is shown, then a parasitic drain exists. To help find the parasitic drain, remove the fuses one at a time until **amps draw** falls to zero. This will help you locate the circuit containing the drain. Once you have identified the circuit, disconnect the components in that system until the current draw falls once again. You can now replace the faulty component.

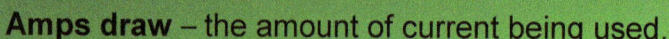

Amps draw – the amount of current being used.

- ✓ Which electrical diagnostic tool can be used to conduct a quick check for low voltage polarity?
- ✓ Apart from voltage, amperage and resistance testing, list three other functions that often available on a multimeter.
- ✓ List five functions available on a typical diagnostic scan tool.

Preparing for assessment

The information contained in this book can help you with theory or practical assessments used to identify your skills or competence when undertaking vehicle repairs or a recognised qualification. It is possible that some of the evidence you produce may contribute to more than one qualification. You should ensure that you make best use of all your evidence to maximise the opportunities for cross-referencing between units or qualifications.

You should choose the type of evidence that will be best suited to the type of assessment that you are undertaking (either theory or practical). The types of evidence you could use are listed below:

- **Direct observation by a qualified assessor**
- **Witness testimony**
- **Computer based**
- **Audio recording**
- **Video recording**
- **Photographic recording**
- **Professional discussion**
- **Oral questioning**
- **Personal statement**
- **Competence/Skills tests**
- **Written tests**
- **Multiple-choice tests**
- **Assignments/Projects**

Before you attempt a written or multiple-choice test, make sure you have reviewed and revised any key terms that relate to the topics in that subject area. Make sure you read all questions carefully. Take time to digest the information so that you are confident about what the question is asking you. With multiple-choice tests, it is very important that you read all of the answers carefully, as it is common for two of the answers to be very similar, which may lead to confusion.

For practical assessments, it is important that you have had enough practice and that you feel that you are capable of passing. It is best to have a plan of action and work method that will help you.

Make sure that you have the correct technical information, in the way of vehicle data, and appropriate tools and equipment. It is also a good idea to check your work at regular intervals. This will help you to be sure that you are working correctly and to avoid problems developing as you work. When undertaking any practical assessment, always make sure that you are working safely throughout the test.

Chapter 2 Hybrid and Electric Vehicles

This chapter gives an overview of the construction and operation of hybrid and electric powered vehicles. It contains procedures you can use when removing, replacing and testing high voltage components to ensure correct safety measures are observed. It will also help you develop a systematic approach to the inspection and maintenance of hybrid and electric vehicles.

Contents

Safe working	65
Information sources	66
Electrical and electronic components	67
Tooling	69
Working perspective	71
High voltage hazards	73
Electrically propelled vehicle types	77
Hybrid drive	79
Regenerative braking	84
Batteries	85
Electric motors	94
The construction and function of associated hybrid components	101
High voltage safety precautions	113
Hybrid internal combustion engines	116
Aerodynamics	121
Transmission	121
Air conditioning	124
Making recommendations	127

Safe working when carrying out maintenance and repairs on hybrid and electric powered vehicles

There are many hazards associated with the inspection and maintenance of vehicle high voltage systems. You should always assess the risks involved with any maintenance or repair routine before you begin and put safety measures in place. You need to give special consideration to the possibility of:

- electrocution from high voltage systems

- coming into contact with chemicals from damaged battery units

Never attempt to work on a high voltage electrical system, unless you have had adequate training.

You should always use the correct personal protective equipment (PPE) when you work on these systems.

When maintaining hybrid and electric drive systems, take care working with the high voltage components. The high voltage system can normally be identified by its reinforced insulation and shielding which is often coloured bright orange. These systems carry voltages that can cause severe injury or death.

Always use the correct tools and equipment. Damage to components, tools or personal injury could occur if the wrong tool is used or misused. Check tools and equipment before each use.

If you are using electrical measuring equipment, always check that it is accurate and calibrated before you take any readings.

If you need to replace any electrical or electronic components, always check that the quality meets the original equipment manufacturer (OEM) specifications. (If the vehicle is under warranty, inferior parts or deliberate modification might make the warranty invalid. Also, if parts of an inferior quality are fitted, they might affect vehicle performance and safety). You should only carry out the replacement of electrical components if the parts comply with the legal requirements for road use.

Information sources

The complex and high risk nature of hybrid vehicle drive systems requires you to have a good source of technical information and data. In order to conduct maintenance and repair procedures, you need to gather as much information as possible before you start. Sources of information include:

Table 2.1 Information Sources

Information source	Example
Verbal information from the driver	A history of the fuel economy and performance output when driving their hybrid car.
Vehicle identification numbers	Vehicle build date and transmission type.
Service and repair history	Has the vehicle been previously repaired with a similar fault.
Warranty information	Is the service history up to date and are the repairs covered under warranty.
Vehicle handbook	How to carry out the charging of the vehicles plug-in electric batteries.
Technical data manuals	The type and quantity of PAG oil to use in a hybrid vehicles air conditioning system.
Workshop manuals	How to remove and refit the electric drive motor from a hybrid car.
Safety recall sheets	A description of a fault and the repair/replacement procedures for the braking systems stroke simulator.
Manufacturer specific	Diagrams showing the wiring layout for a vehicles high voltage system.
Information bulletins	A description of a vehicle modification to improve the air cooling of the high voltage battery unit.
Technical help lines	Advice on specialist tools required for the removal and refitting of the motor/generator stator unit in a hybrid drive system.
Advice from master technicians/colleagues	A demonstration of how to safely isolate the high voltage system on a hybrid drive car.

Table 2.1 Information Sources

Internet	An Internet forum page showing where a number of electric vehicle owners are having similar problems with the driving range when the car is used on a motorway.
Parts suppliers/catalogues	Cross reference numbers showing that the brake pads supplied for a hybrid car are compatible with its braking system.
Jobcards	A description of the work to be carried out; service type for example.
Diagnostic trouble codes	A four digit area trouble code, followed by a specific three digit code. For example: P1600 Battery malfunction, 117 High voltage ECU back-up power source circuit malfunction.
Oscilloscope waveforms	The waveform generated by a wheel speed sensor used in conjunction with the brake-by-wire system.
On vehicle warning labels/stickers	Vehicle safety warning labels identifying the high voltage system components under the bonnet.

Always compare the results of any inspection or testing to suitable sources of data. Remember that no matter which information or data source you use, it is important to evaluate how useful and reliable it will be to your safety, maintenance and repair routine.

Table 2.2 The operation of electrical and electronic systems and components related to hybrid and electric drive systems

Electrical/Electronic system component	Purpose
ECU	The electronic control unit is designed to monitor the operation of vehicle systems. It processes the information received and operates actuators that control system functions. An ECU may also be known as an ECM (electronic control module).
Sensors	The sensors are mounted on various system components and they monitor the operation against set parameters. As the vehicle is driven, dynamic operation creates signals in the form of resistance changes or voltage which are sent to the ECU for processing.

Table 2.2 The operation of electrical and electronic systems and components related to hybrid and electric drive systems

Actuators	When actioned by the ECU, motors, solenoids, valves, etc. help to control the function of the vehicle for correct operation.
Digital principles	Many vehicle sensors create analogue signals (a rising or falling voltage). The ECU is a computer and needs to have these signals converted into a digital format (on and off) before they can be processed. This can be done using a component called a pulse shaper or Schmitt trigger.
Duty cycle and PWM	Lots of electrical equipment and electronic actuators can be controlled by duty cycle or pulse width modulation (PWM). These work by switching components on and off very quickly so that they only receive part of the current/voltage available. Depending on the reaction time of the component being switched and how long power is supplied, variable control is achieved. This is more efficient than using resistors to control the current/voltage in a circuit. Resistors waste electrical energy as heat, whereas duty cycle and PWM operate with almost no loss of power.
Networking and multiplex systems	Many modern vehicle systems are controlled using computer networking. In these systems a number of ECU's are linked together and communicate to share information in a standardised format. The most common network system is Controller Area Network (CAN Bus).
High voltage rectification	The high voltage batteries used in hybrid and electric vehicles work with direct current DC. The charging systems for these batteries will normally supply an alternating current AC (including mains plug-in). Before AC electricity can be used to charge the batteries, it must be converted to DC and this is done via a unit known as a rectifier. The rectifier contains a number of power transistors and diodes that redirect the electric current to enable it to be used in the batteries. This high voltage DC can then be used to charge the battery of the low voltage (12 volt) system if it is put through a step-down transformer known as a DC to DC converter.
High voltage motor inversion	The high voltage electric drive motors used in hybrid and electric vehicles normally use alternating current AC. In order to use the electricity stored in the batteries, the direct current must be converted into alternating current. This is done with an inverter which uses transistors to oscillate the electric power.

Tooling

Table 2.3 shows a suggested list of diagnostic and maintenance tooling that could be used when testing and evaluating hybrid or electric vehicle drive systems. Due to the nature of high voltage systems, you will experience different requirements during your diagnostic and repair routines, therefore you will need to adapt the list shown for your particular situation.

Table 2.3 Tooling

Tool	Possible use
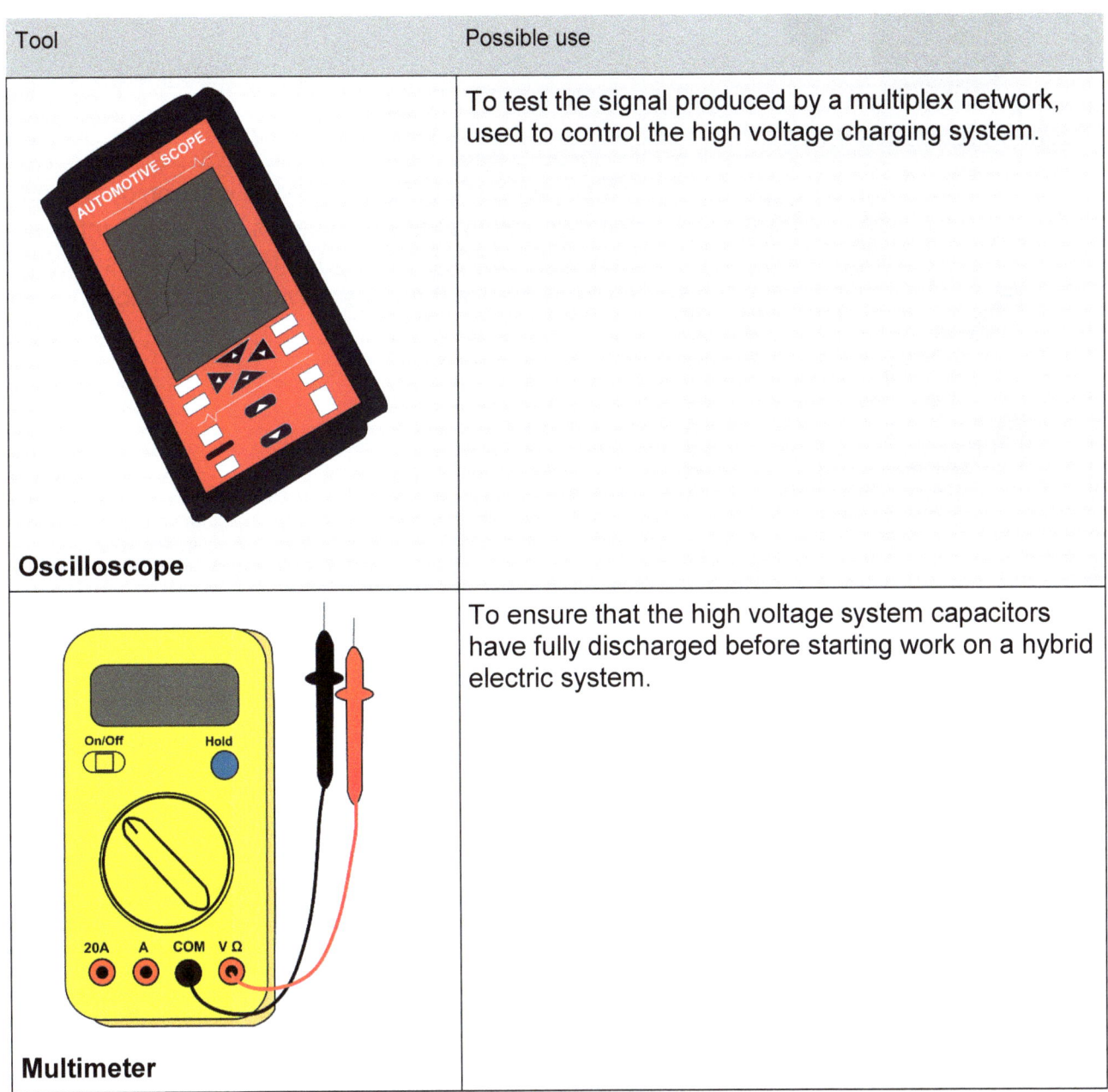 **Oscilloscope**	To test the signal produced by a multiplex network, used to control the high voltage charging system.
Multimeter	To ensure that the high voltage system capacitors have fully discharged before starting work on a hybrid electric system.

Table 2.3 Tooling

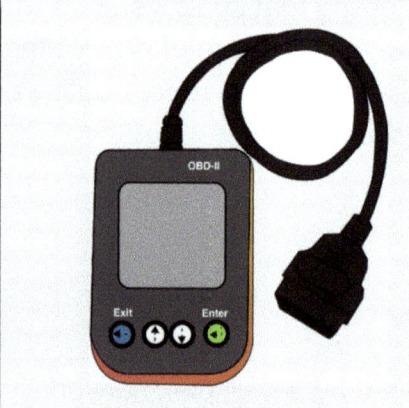 **Code reader/scan tool**	To retrieve diagnostic trouble codes (DTC) related to the hybrid drive high voltage system. To retrieve live data, clear trouble codes, reset the malfunction indicator lamp, and evaluate the effectiveness of repairs.
 **Hand tools**	Whenever hand tools are used to work on or around the high voltage vehicle systems, it is important that they are specifically designed to be highly insulated so as to reduce the possibility of electric shock.
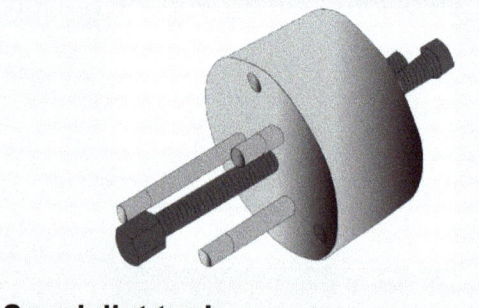 **Specialist tools**	A puller may be required to remove and refit the rotor of a hybrid drive electric motor, to reduce the possibility of damage or injury.

All tools and equipment used when working on hybrid and electric vehicles should be accurately calibrated before each use and fit for purpose.

Working perspective

Working on or around vehicles with alternative propulsion systems presents a unique set of issues depending on your perspective. A number of approaches may be required for individual situations relating to the type of level and involvement you have with a given scenario.

Your involvement might include:

- Being a member of the emergency services, such as police, fire, ambulance or coastguard

- Being involved in the motor industry, such as recovery operator, vehicle technician or end of life vehicle recycling

- Being involved in training, such as lecturer, workshop instructor, assessor or workplace mentor

The following activity describes a scenario relating to alternative propulsion vehicles and lists a set of suggested precautions or actions that could be undertaken, see table 2.4. These actions are not in any particular order or restricted to an individual profession, perspective or approach.

Using the information provided in this chapter and the knowledge of your own role or position, create a list of actions that you would take in this situation.

Try to put these in a logical order and include any extra steps that you feel are missing or are unique to your role or perspective. (You can also attempt to make lists from the perspectives of other professions or rolls from the lists above).

When complete, share your list with colleagues or peers to compare your results.

Table 2.4 Working Perspective

Scenario	Possible steps
• You are preparing to work on a hybrid electric vehicle at the roadside or in a workshop. • The work could be minor, and may involve the risk of coming into contact with high voltage systems or the engine starting automatically when not expected. • The work could be major involving the high risks associated with a vehicle that has become badly contaminated with water (i.e. submerged).	• Isolate the electrics • Have a suitable fire extinguisher handy • Ensure that the smart key is beyond its range of operation • Wear electrically insulated gloves • Ensure that others are aware of the situation • Do not work around the strong magnetic fields of motor/generators with a heart pacemaker • Conduct a risk assessment • Remove all sources of ignition • Allow time for system capacitors to discharge • Wear high visibility clothing • Allow any water to drain naturally from the vehicle • Remove electrical system fuses • Wear eye protection • Wear overalls with non-conducting fasteners • Wear insulating/Wellington boots • Secure the vehicle to prevent it moving unexpectedly • Disable the fuel injection system • Disable the ignition system • Remove the owners belongings • Measure the voltage at the capacitors • Isolate SRS airbags • Prevent accidental reconnection of the high voltage system

Table 2.5 Hazards associated with high voltage electrical systems found on vehicles

Hazard	Risk or danger
Electric shock	Electric shock is caused when electricity passes through the human body leading to injury or death. The risk of electric shock increases as electrical voltage rises. Voltages higher than 60 volts DC or 30 volts AC rms values increase the risks posed.
Burns	A common effect of electric shock on the human body is burning. This is due to the fact that the electrical energy is converted to heat as it is discharged. Burns caused by electrical discharge through the body may be internal and cause tissue damage that might not be immediately apparent. If an accidental short circuit is created during the connection and disconnection of a high voltage electrical circuit, the heat given off can cause external burning to the skin.
Arc flash	Arc flash is caused by a sudden, accidental discharge of electrical energy and creates an arc similar to that of welding. The temperatures created when this occurs can be around 20,000 degrees Celsius. This temperature is enough to vaporise clothing and human flesh. Arc flash is most likely to occur at voltages above 480 volts, which are not uncommon on electric vehicle drive systems.
Arc blast	Arc blast is an explosion caused by a short circuit in a high voltage system. It has many of the same effects as arc flash, causing severe burns, but also has the potential to cause injury at a distance. Added dangers may involve flying particles from the system, and a sound pressure wave that might lead to permanent hearing damage.
Fire	A rapid discharge of electrical energy will create large amounts of heat. This heat will then have a high potential of igniting flammable materials. When working on high voltage systems it is advisable to have a suitable fire extinguisher to hand (one that will not conduct electricity - class E).
Explosion	High voltage systems often combine electricity with chemicals, gasses and fumes that pose a risk of explosion. Care should always be taken to try and remove any sources of ignition and correctly isolate electrics before working on these systems.

Levels of current and voltage that may present hazards

An electric shock is usually painful and can be lethal. The level of voltage is not a direct guide to the level of injury or danger of death. Physical effects and damage are generally determined by the amount of current and the duration of electric shock. Even a low voltage causing a current of extended duration can be fatal.

The type of damage caused by electric shock can be different depending on the voltage, duration, current and path taken.

- Current entering the hand has a threshold of around 5 to 10 milliamps (mA) for DC ⎓ and about 1 to 10 mA for AC ∿.

- At Voltages of between 110 to 220 volts AC ∿, current traveling through the chest for a fraction of a second can cause a heart attack with currents as low as 60mA.

- With DC ⎓, 300 to 500 mA is required.

For further guidance on the dangers posed by electric current, see figure 2.1.

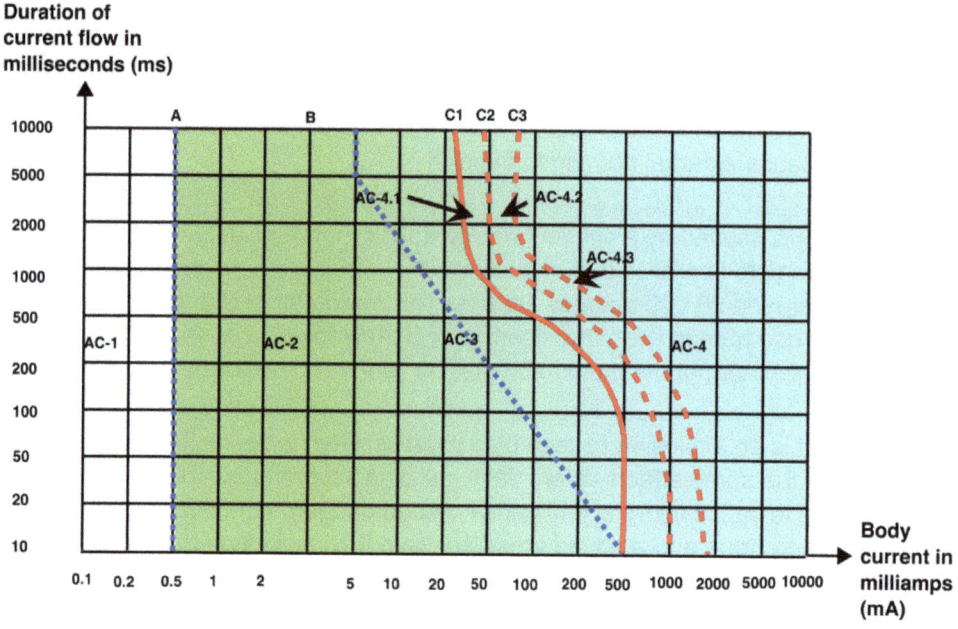

Figure 2.1 The dangers posed by electric current

AC-1 zone: Imperceptible

AC-2 zone: Perceptible

AC-3 zone : Reversible effects: muscular contraction

AC-4 zone: Possibility of irreversible effects

AC-4-1 zone: Up to 5% probability of **heart fibrillation**

AC-4-2 zone: Up to 50% probability of heart fibrillation

AC-4-3 zone: More than 50% probability of heart fibrillation

A curve: Threshold of perception of current

B curve: Threshold of muscular reactions

C1 curve: Threshold of 0% probability of **ventricular fibrillation**

C2 curve: Threshold of 5% probability of ventricular fibrillation

C3 curve: Threshold of 50% probability of ventricular fibrillation

Hybrid Electric & Alternative Automotive Propulsion

Notes: The effects of electric current and the standards needed to protect operators from the dangers they pose are regulated by the International Electrotechnical Commission - ICE 60479 and ICE 479-2.

Terms:

Heart fibrillation - a heart condition that causes an irregular and often abnormally fast heart rate.

Ventricular fibrillation - a condition in which there is uncoordinated contraction of the cardiac muscle of the ventricles in the heart, making them quiver rather than contract properly.

Notes: Most of the high voltage wiring is covered in brightly coloured orange insulation and contained within reinforced housings and enclosures which are mounted in locations to keep them out of contact with users or operators.

Manufacturers of high voltage vehicle systems will include a number of safety devices to help protect the operators conducting maintenance and repairs. They will also help reduce damage to vehicle systems and components. Safety devices may include:

- Fuses
- RCD - Residual Current Devices (a form of circuit breaker)
- RCBO - Residual Current Circuit Breaker with Overload protection
- MCB - Miniature Circuit Breaker

These devices are designed to make the systems open circuit if a fault occurs which draws excessive current. Once the system has become open circuit, no more current flows and further damage or injury is prevented.

Low Carbon Technologies

In the event of an accident, hybrid and electric vehicle high voltage systems may be exposed. This has the potential to cause harm to:

- occupants
- on-lookers
- recovery personnel
- emergency services

If you are attending an incident involving a hybrid or electrically propelled vehicle you must assess the risks involved before taking any action.

Isolate high voltage components

Before any work is started on high voltage systems, the circuits should be isolated.

How to:

1. Ensure that you are wearing the correct PPE.
2. Remove the key from the ignition.
3. If the ignition key is a 'smart key', make sure that it is stored away from the car, outside its range of operation.
4. Following manufacturer's instructions and any system warning labels, locate the high voltage battery isolation switch/plug.
5. Switch off high voltage isolation switch or remove the plug and apply any safety catches that are available in order to prevent accidental reconnection.
6. Locate the high voltage system output terminals, and connect a multimeter set to 400 volts DC ⎓, to measure the system voltage. (Ensure your fingers stay behind the insulated finger guards of the voltage probes).
7. Allow the system capacitors to discharge and do not start any further work until voltages have fallen to the manufacturer recommended levels. (If possible, wait until system voltage has fallen to 0 volts).

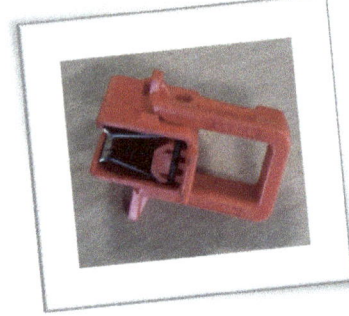

Figure 2.2 A high voltage isolation plug

Hybrid Electric & Alternative Automotive Propulsion

Examples of the electrically propelled and hybrid vehicles that are currently available

Figure 2.3 Two wheel moped/scooters

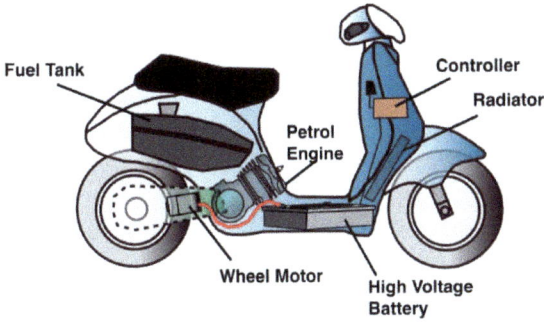

Figure 2.4 Cars

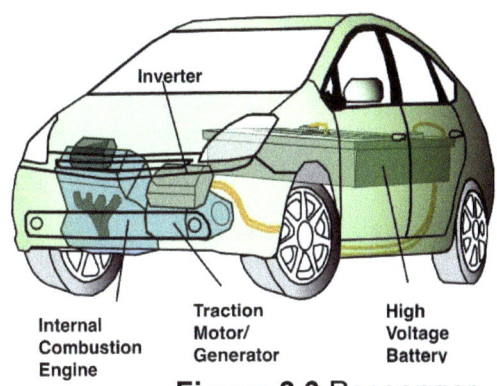

Figure 2.5 Commercial vehicles

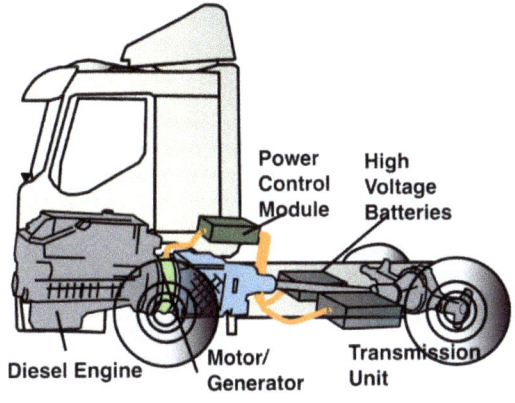

Figure 2.6 Passenger transport

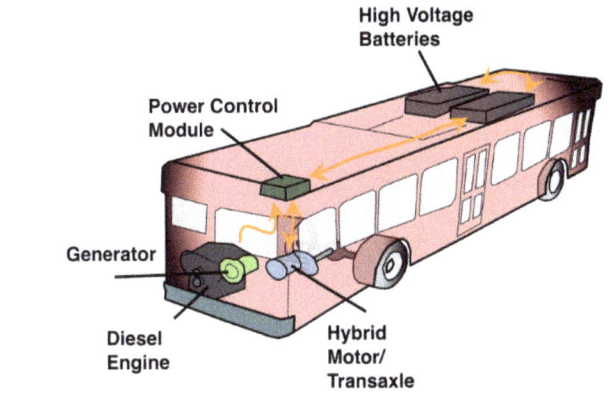

All electric vehicles (AEV)

An all-electric vehicle is one where its main source of power for propulsion comes from high voltage batteries and motors. These types of vehicles are sometimes known as battery electric vehicles or BEV's, and because they are often charged from mains electricity they are also called 'plug-in'.

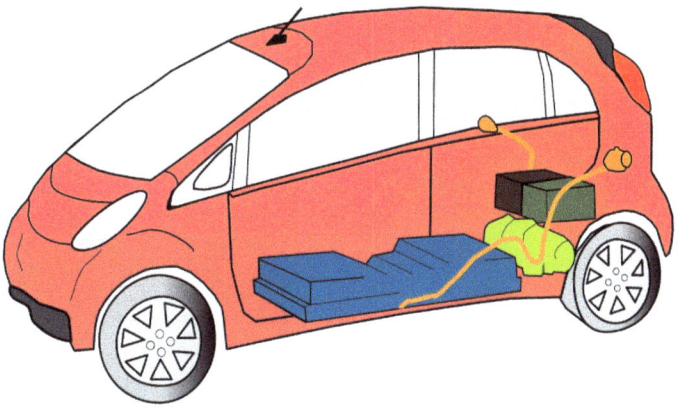

Figure 2.7 A battery electric 'plug-in' vehicle

- List three hazards associated with vehicle high voltage systems.
- With hybrid and electric drive vehicles, in the event of an accident, who might be harmed?
- What colour is the insulation of high voltage cabling?

Hybrid vehicles

A hybrid vehicle is one where the main source of power for propulsion is provided by a combination of internal combustion engine and electric **traction motor**. The engine and electric motor can be connected in three main formats to provide drive:

Series

Parallel

Combination (series parallel)

Many modern hybrid vehicles are of the parallel or combination type and are made up of the following components:

- batteries
- traction motors/generators
- cabling
- control units
- circuit protection

Traction motor - a powerful electric motor used to provide the driving force in hybrid and electric vehicles.

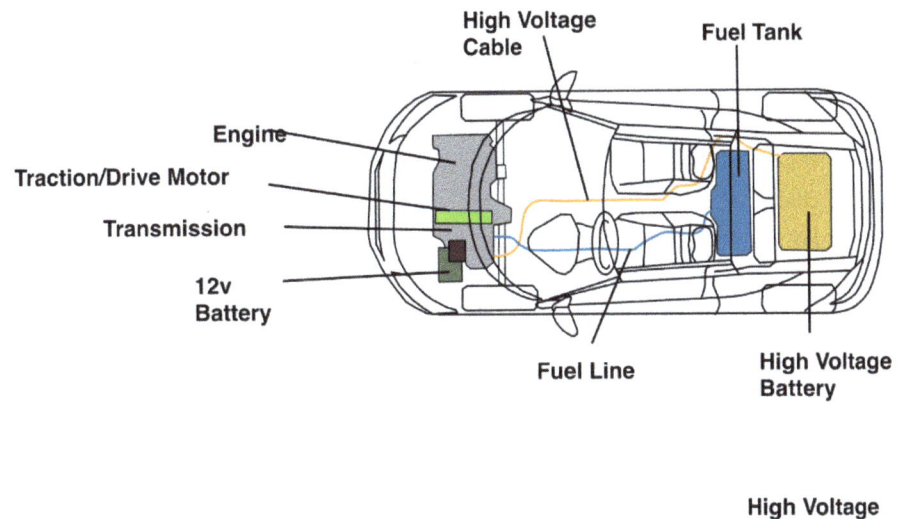

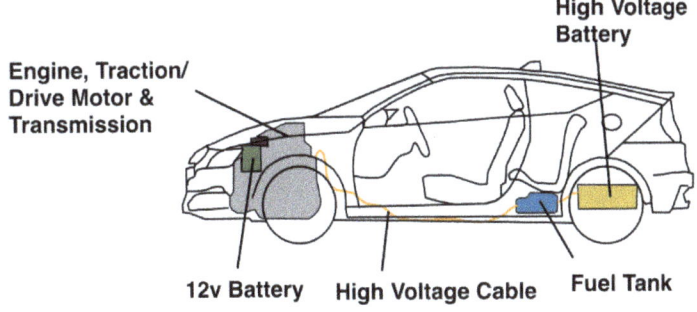

Figure 2.8 A hybrid vehicle system layout

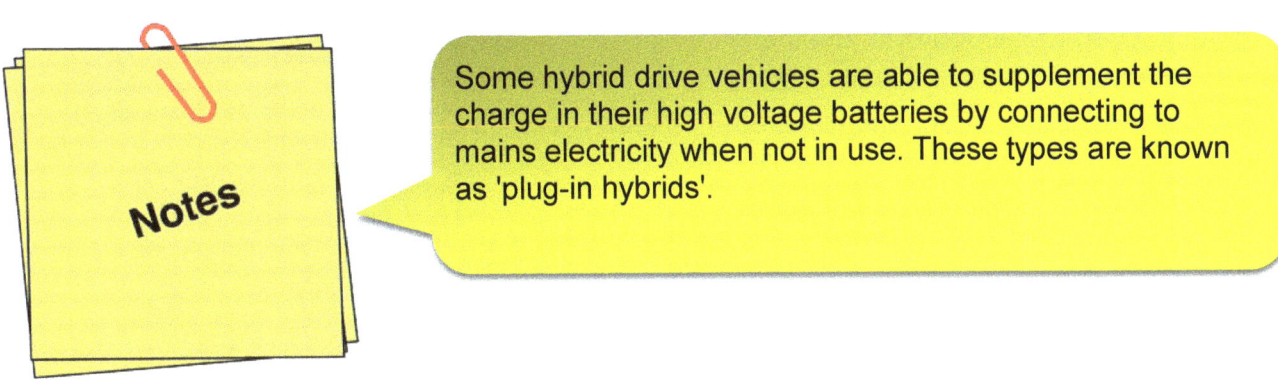

> Some hybrid drive vehicles are able to supplement the charge in their high voltage batteries by connecting to mains electricity when not in use. These types are known as 'plug-in hybrids'.

Hybrid drive

A hybrid vehicle is one which combines an internal combustion engine with an electric motor to provide drive. This gives the flexibility of a petrol or Diesel engine with the fuel economy and low pollution characteristics of electric motors.

There are three main types of hybrid drive:

- **Series hybrid:** A small capacity internal combustion engine is used to act as a generator. This then charges batteries that are used to power the electric motors that drive the wheels. There is no direct connection between the engine and the wheels, meaning that a gearbox is not required. The advantage of this system is that no driving loads are placed on the engine and it can run at a constant speed. This reduces fuel consumption, emission output and engine wear.

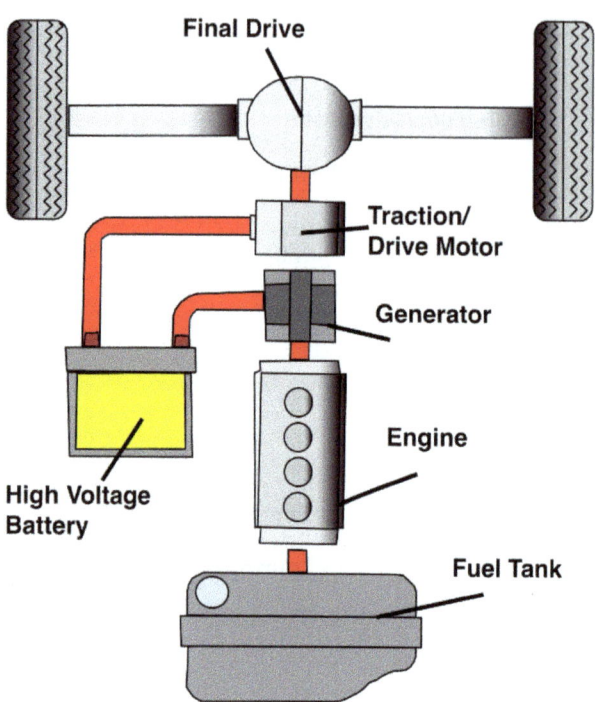

Figure 2.9 Series hybrid drive

- **Parallel hybrid:** An integrated electric motor is used to support or boost the performance of a small capacity internal combustion engine. When not required, the electric motor can be converted into a generator to recharge the high voltage electric batteries.

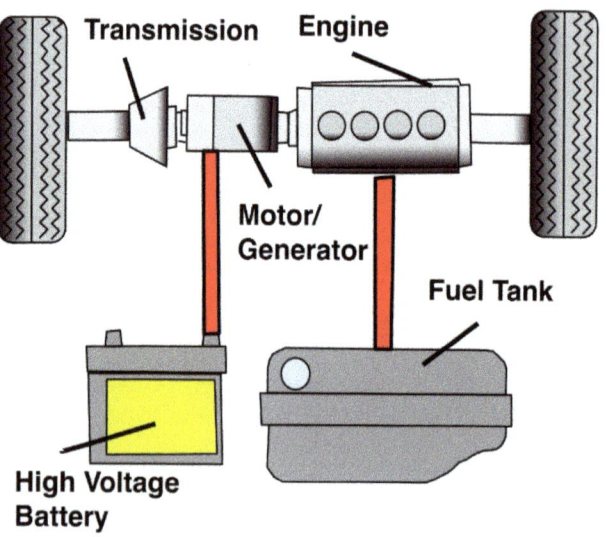

Figure 2.10 Parallel hybrid drive

Hybrid Electric & Alternative Automotive Propulsion

- **Combination hybrid:** This type of hybrid uses the properties of both series and parallel hybrids. The car can operate on electric motors alone, internal combustion engines alone or a combination of both.

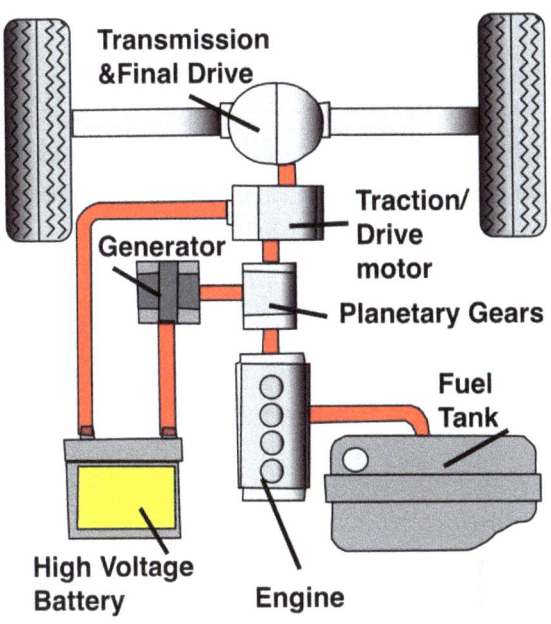

Figure 2.11 Combination hybrid drive

Vehicle designers are continuously creating unusual types of hybrid drive systems. This not only improves efficiency and operation, but also overcomes **copyright** issues.

Some manufacturers have been known to use the terms **series** and **parallel** in a slightly different context when describing the operation of their hybrid drive systems. Make sure that you have fully studied and understood the manufacturer's description of their system before you make any assumptions about how they operate. This will help reduce confusion and misunderstanding.

Copyright – an exclusive legal right of design ownership and use.

Series – connected in a line (one after another).

Parallel – connected side by side.

Two common hybrid systems are described in the next section.

Collaborative Motor Drive

A **Collaborative** Motor Drive system has two separate motors that run parallel and are capable of producing drive: an internal combustion engine and an electric traction motor. When required they work together to provide a smooth drive system.

- When pulling away, the petrol engine is not used and the electric motor provides the drive.

- When going up a hill, the engine and electric motor both power the vehicle to provide maximum performance.

- When decelerating or braking, the system recycles the kinetic energy to recharge the batteries.

- When overtaking, the engine and electric motor both power the vehicle to provide maximum performance.

- When the vehicle is stationary, both the engine and the electric motor automatically switch off to save power.

Integrated Motor System

An Integrated Motor System has a compact electric motor, sandwiched between an internal combustion engine and the gearbox, in a similar position to the flywheel. When needed this electric motor is able to boost, the performance from the engine.

- When pulling away, the electric motor provides maximum torque to assist the engine for strong acceleration and reduced fuel consumption.

- When the vehicle is cruising at low speed, the engine is stopped (intake and exhaust valves remain closed) and the electric motor is used by itself for drive.

- When the vehicle is accelerating gently or cruising at high speed, the vehicle runs on engine power alone and the electric motor is switched off.

- When accelerating rapidly, the engine and electric motor both power the vehicle to provide maximum performance.

- When decelerating or braking, the engine is stopped (intake and exhaust valves remain closed) and the electric motor acts as a generator to recharge the batteries.

- When the vehicle is stationary, the engine and electric motor automatically switch off to save power.

Hybrid Electric & Alternative Automotive Propulsion

Collaboration – working together in co-operation (this is also known as synergy).

Hybrid vehicle regenerative braking

A standard braking system uses friction to convert the kinetic (movement) energy of a vehicle into heat. Hybrid vehicles often use a different method for braking and slowing down, which is called **regenerative** braking. This is a highly efficient process that turns some of the vehicle's kinetic energy into electricity that can be used to help charge the high voltage battery system.

Because a hybrid vehicle uses a combination of an internal combustion engine and electric motor to provide drive, if the electric motor is driven mechanically during braking for example, it can be converted into an electrical generator.

The conversion of kinetic energy into electricity actually slows the vehicle down.

Any extra deceleration required by the driver that is not achieved by the regenerative braking is handled by a brake-by-wire system which operates brake callipers and pads against discs. A sophisticated electronic control system is used to calculate the amount of braking required and splits the operation between the generator and the brakes.

Limitations of regenerative braking include:

- The regenerative braking effect drops off at lower speeds.
- Most road vehicles with regenerative braking only have power on some wheels, for example a two-wheel drive car. The regenerative braking power only applies to the drive wheels because they are the only wheels linked to the drive motor. In order to provide controlled braking under difficult conditions, such as wet roads, friction-based braking is necessary on the other wheels.
- If the batteries or capacitors are fully charged, no regenerative braking takes place.

Figure 2.12 Hybrid vehicle regenerative braking

Regenerative – a method of braking in which energy is extracted from the parts braked, to be stored and reused.

Brake by wire

The regenerative braking used in hybrid vehicles means that only some of the energy needed to slow the vehicle down comes from the hydraulic system. To ensure that the deceleration of the car is accurately controlled, manufacturers incorporate brake-by-wire systems in their vehicle design. Instead of the master cylinder applying hydraulic pressure directly to the brake calliper system, it operates as a pressure measurement sensor (also known as a stroke simulator). This sensor simulates pedal pressure so that brake operation feels normal to the driver. The signal from the brake pressure sensor is processed by an ECU, which operates a secondary master cylinder unit. In conjunction with the wheel speed sensors, the secondary master cylinder applies the appropriate brake force required to slow the wheels without allowing them to skid. If the regenerative braking and hydraulic brake by wire fails, the system enters conventional brake operation, which will allow the vehicle to be slowed down but the driver may have to push the pedal slightly harder and overall stopping distances may be increased.

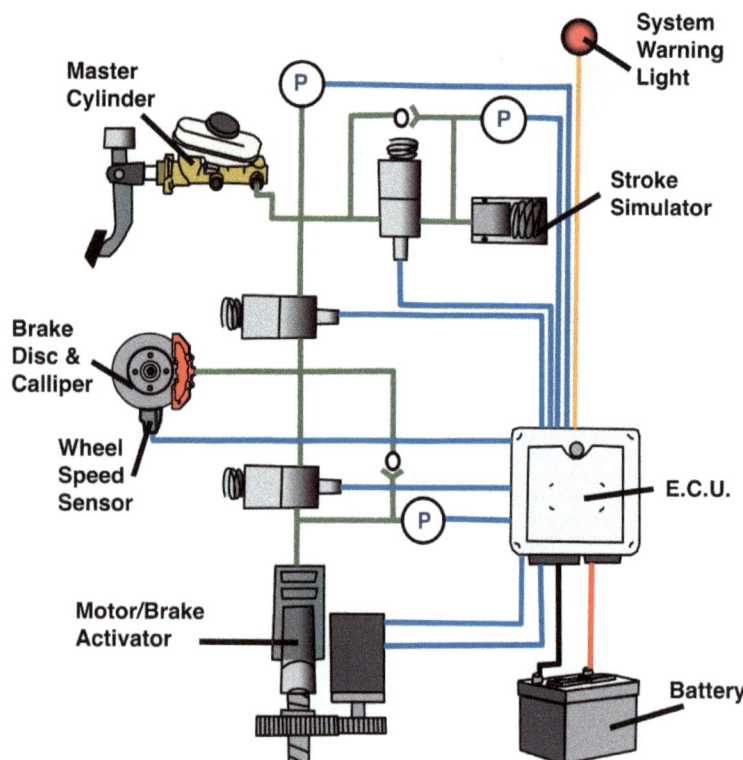

Figure 2.13 Brake by wire

The construction and function of battery modules

Battery types

Low voltage lead acid

A standard lead-acid battery, the type often found in cars, contains a number of sections called cells (see Figure 2.14). Each cell is capable of producing approximately 2.1 volts.

A standard battery contains six cells linked together, creating a battery with a voltage of 12.6 volts. This is often rounded down, so we say that the battery has a voltage of 12 volts.

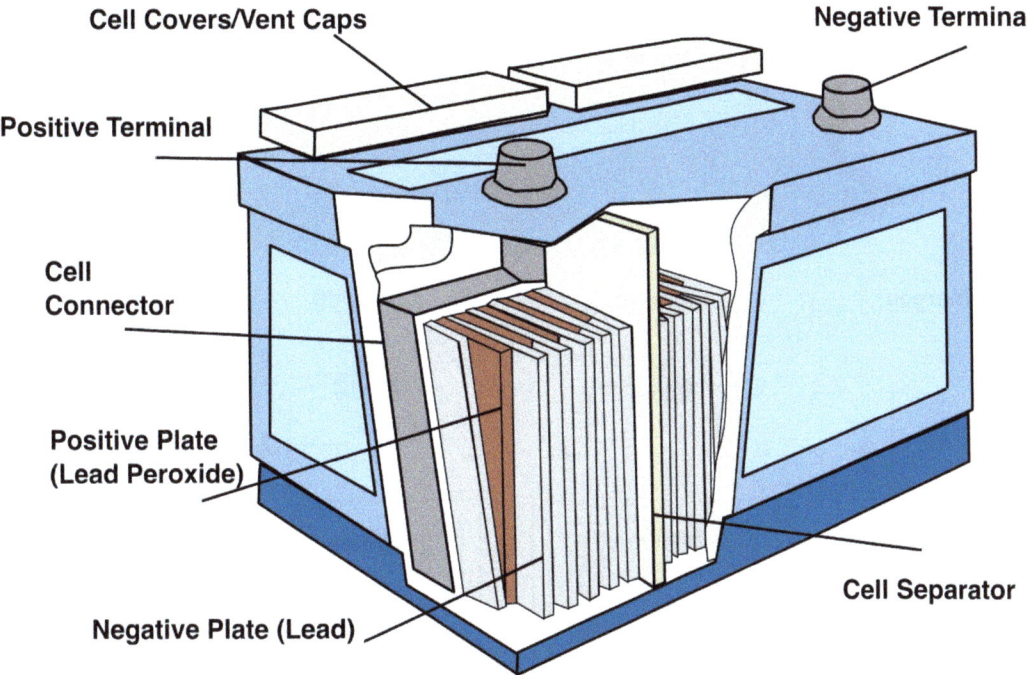

Figure 2.14 The components of a lead acid battery

Each cell contains a number of lead plates, which are chemically different:
- The negative plate is made of lead.
- The positive plate is made of lead peroxide.

To prevent the plates from touching each other and causing a short circuit, thin sheets of material called separators are inserted between them.

Lead peroxide contains extra oxygen compared with normal lead. This means that the positive plate is chemically different from the negative plate, and it would like to share its electrons with the negative plate.

If connected in a circuit, the electrons are allowed to move from the positive plate to the negative plate. This creates electric current and provides the energy to power components in the vehicle.

The first half of the circuit is made with an **electrolyte** which consists of sulphuric acid and deionised water. The liquid covers the plates and allows electrons to move from one plate to another through the electrolyte. The top of each plate is then connected to the rest of the circuit. The circuit must contain a consumer to use up the electrical potential energy.

When the circuit is complete, the electrons combine with the electrolyte and move from one plate to another as a chemical reaction, creating current.

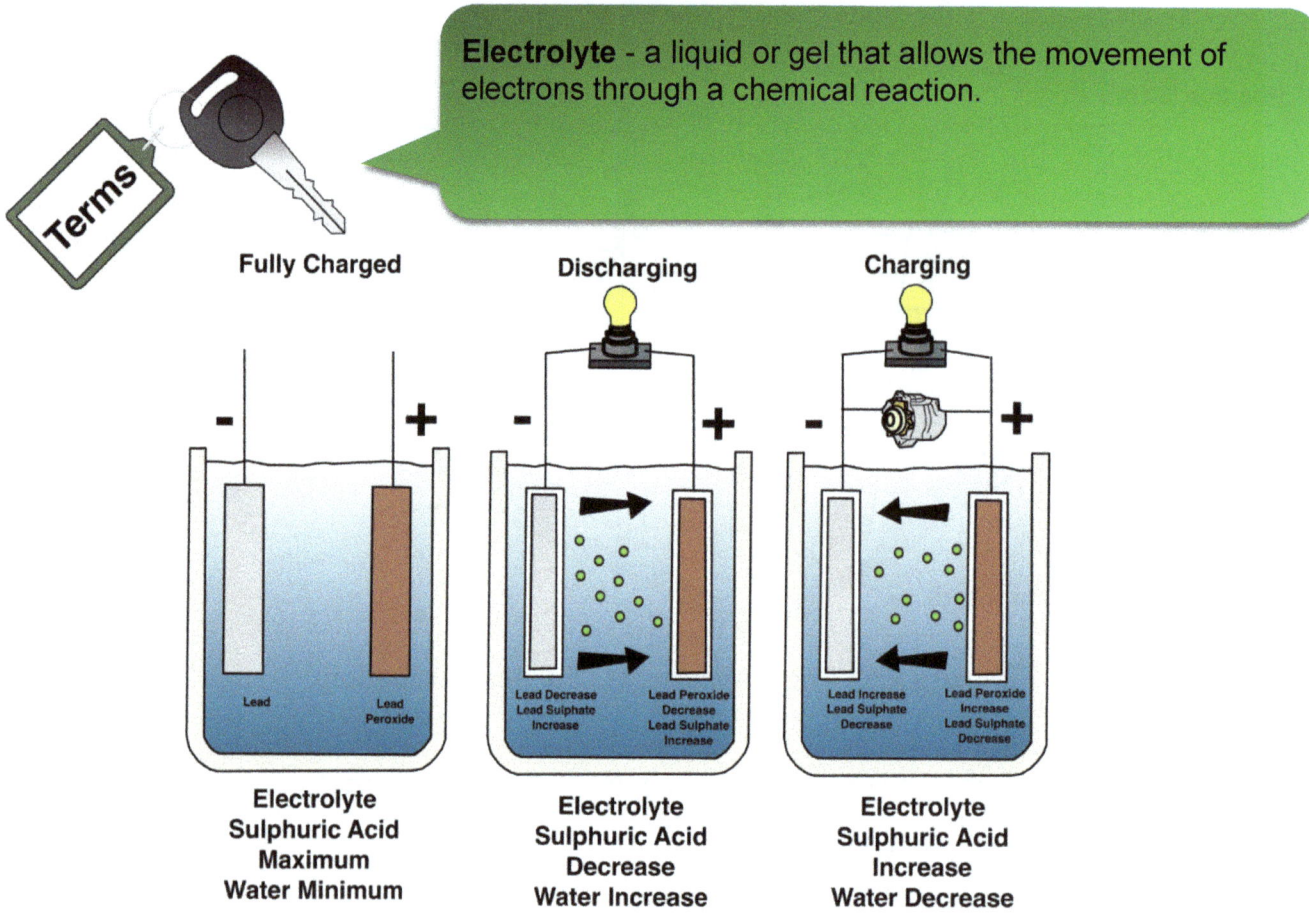

Figure 2.15 The chemical reactions taking place inside in a lead acid battery

Why batteries need to be recharged

Batteries work with electricity flowing in one direction only (direct current). This means that eventually both plates in a low voltage lead acid battery will become chemically the same (they change to a substance called lead sulphate) and current will stop.

When this happens, no more electricity flows through the circuit, meaning that the battery is flat. When the battery is flat, it needs to be recharged. To do this, a generator (which can be thought of as a pump for electricity) is connected to the engine.

To recharge the battery, an electric voltage is supplied to the battery circuit with a pressure that is higher than the EMF (approximately 12.6 volts with all electrical consumers switched off). This reverses the chemical reaction – it forces electrons back through the electrolyte to their original positions in the lead and lead peroxide plates and recharges the battery.

In a hybrid car the high voltage batteries will also need recharging. In these systems, the generator is often the electric drive motor turned in reverse.

The hybrid drive motor will first charge the high voltage batteries which serve the hybrid drive system. The voltage is then stepped down through a DC === to DC === converter, where it is applied to the low voltage battery circuit.

Table 2.6 explains some terms associated with lead acid batteries and some of their classifications/ratings.

Table 2.6 Battery terms and ratings

Term/Rating	Description
Amp hours (Ah)	A measurement of the electrical current that a battery can deliver. This quantity is one indicator of the total amount of charge that a battery is able to store and deliver at its rated voltage. The amp hours value is the total of the discharge-current (in amperes), multiplied by the duration (in hours) for which this discharge-current can be sustained by the battery. For example, a car battery could be rated as 100 Ah, which should contain enough electricity to provide: 100 amps for one hour 1 amp for 100 hours 10 amps for 10 hours Any other combination that multiply together to make 100 (e.g. 25 amps for 4 hours). The amp hour rating, is required by law in Europe, to be shown on a battery.

Table 2.6 Battery terms and ratings

Cranking amps (CA)	A number that represents the amount of current a lead-acid battery can provide at 0°C (32°F) for 30 seconds and maintain at least 1.2 volts per cell (7.2 volts for a 12 volt battery).
Cold cranking amps (CCA)	A number that represents the amount of current a lead-acid battery can provide at −18°C (0°F) for 30 seconds and maintain at least 1.2 volts per cell (7.2 volts for a 12 volt battery). This test is more demanding than those conducted at higher temperatures.
Hot cranking amps (HCA)	A number that represents the amount of current a lead-acid battery can provide at 27°C (80°F) for 30 seconds and maintain at least 1.2 volts per cell (7.2 volts for a 12 volt battery).
Reserve capacity minutes (RCM), also known as reserve capacity (RC)	A lead-acid battery's ability to sustain a minimum stated electrical load. It is defined as the time (in minutes) that the battery at 27°C (80°F) will continuously deliver 25 amperes before its voltage drops below 10.5 volts.

Electronic car battery testers are now available. When they are connected to a battery and programmed with details found on the battery casing, they run through a series of checks and display the results of the analysis on a screen.

Absorbed Glass Matt (AGM) low voltage battery

Some manufacturers use a form of lead acid battery to power the low voltage vehicle systems which is known as an Absorbed Glass Matt (AGM) battery. In this type the electrolyte is held in a glass fibre mesh which acts as separators between the plates. By containing the electrolyte in the glass matt, the amount of hydrogen gas given off during charging is considerably reduced. The battery is vented to external air to reduce any excessive gas build-up during operation and ensure that any internal pressures are kept within acceptable limits.

Due to their construction, AGM batteries need to have their charging carefully controlled as excess voltage and current can easily damage the internal components. If external charging is required, always use an approved battery charger.

Because the electrolyte of an AGM battery is not in a free liquid state, it cannot be topped up, so the battery is sealed for life in a maintenance free (MF) form.

Nickel cadmium battery

An alternative to the lead-acid battery is the nickel cadmium (NiCad) type, which is found in some cars. It works in a similar way to a standard lead acid battery but requires less maintenance and cannot be overcharged. This type of battery is made of the following materials:

- positive plates: nickel hydrate
- negative plates: cadmium
- electrolyte: potassium hydroxide and water

These batteries tend to be larger and more expensive than normal lead-acid batteries. However, they are better at coping with the extreme loads placed on them by modern electrical systems, especially in hybrid or battery electric vehicles.

Nickel-Metal Hydride battery (Ni-MH)

Nickel-Metal Hydride batteries are becoming the most popular for use with hybrid drive vehicles. They are very similar in construction to a nickel cadmium battery, but use a metal hydride (hydrogen atoms stored in metal) as the negative plate. A Ni-MH battery can have two or three times the capacity of an equivalent sized nickel cadmium battery, meaning physical size is reduced but electrical storage is increased. A single cell of a Nickel-Metal Hydride battery will produce 1.2 volts and are connected in series to each other in groups of six to form packs known as modules. The modules are then connected in series to form the complete high voltage battery.

Current hybrid drive systems (depending on manufacturer) are using batteries with a high voltage potential somewhere between 100V and 300V.

A Nickel-Metal hydride battery has the following advantages over other battery types:

- They have a high electrolyte conductivity, which allows them to be used in high power applications (such as hybrid drive and electric vehicles).
- The battery system can be sealed, which minimizes maintenance and leakage issues.
- They operate over a very wide temperature range.
- They have very long life characteristics when compared with other battery types – this offsets their higher initial cost.
- They have a higher energy density and lower cost per watt than other battery types.

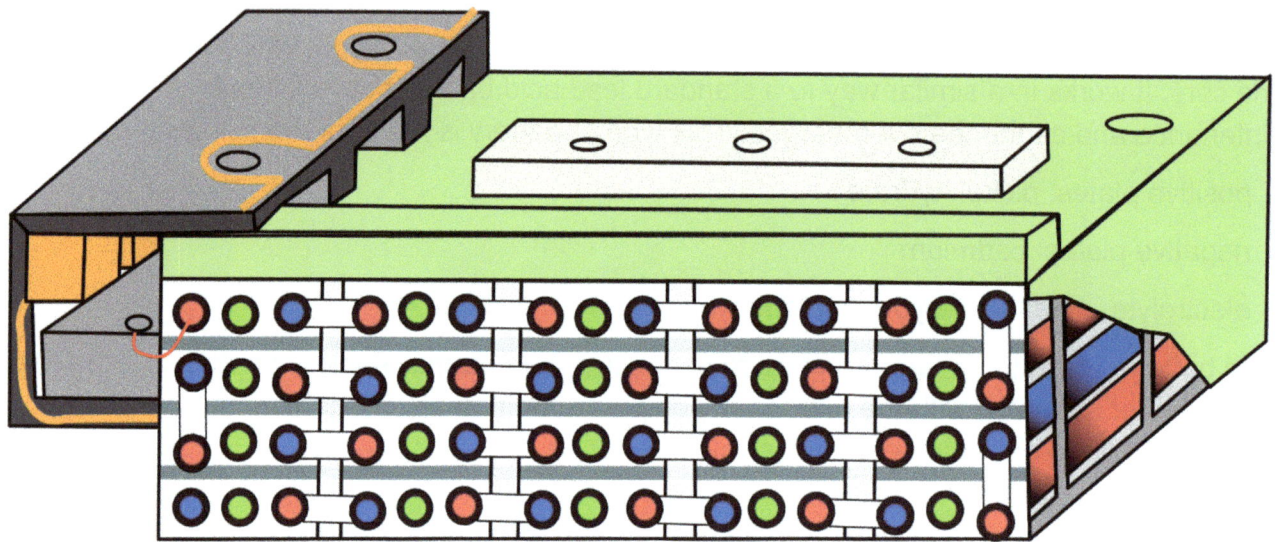

Figure 2.16 Nickel-Metal Hydride battery (Ni-MH)

Because of the characteristics of a Nickel-Metal hydride battery, the charging and discharging has to be very carefully monitored and controlled. If the battery pack is allowed to charge too quickly, overheating and damage can occur. The hybrid drive generator system will maintain a constant current to **trickle charge** the high voltage battery to try and maintain a set **state of charge (SOC)**. An ideal state of charge is around 60% of battery capacity, allowing the battery to work well within its capabilities.

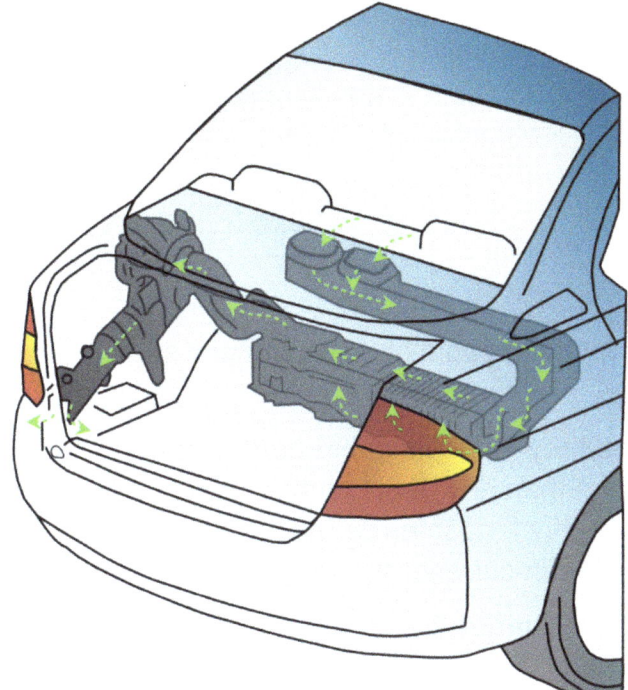

Battery temperature is carefully monitored because as the cells become fully charged any excess energy will be converted into heat. **Thermistors** are used to measure battery temperature and if required an electronically controlled fan is able to draw air through ducting surrounding the high voltage battery unit and assist with cooling.

Figure 2.17 A high voltage battery cooling unit

Trickle charge - a slow charging method that is equal to or very slightly above the batteries natural discharge rate.

State of charge SOC - a rating that shows how much electricity is contained in the battery compared to its capacity.

Thermistors - temperature sensitive resistors.

The measurement of temperature in the high voltage battery system is also a good indication of the battery's state of charge (SOC). By monitoring temperature, the battery system ECU is able to send signals to the generator regulation control unit and alter the amount of charge supplied.

Lithium Ion Battery

For future hybrid and electric cars, a better battery type capable of delivering and storing high voltage electricity is needed.

Lithium ion (Li-ion) batteries are important because they have a higher energy density, (the amount of energy they hold by volume, or by weight) than any of the other currently used batteries. Generally, the Li-ion battery cells hold roughly four times as much energy as the Nickel-Metal Hydride (NiMH) batteries which are used in current hybrids.

Remove and refit high voltage system component (battery)

How to:
1. Ensure that you are wearing the correct PPE.
2. Remove the key from the ignition
3. If the ignition key is a 'smart key', make sure that it is stored away from the car, outside its range of operation.
4. Isolate the high voltage system, ensuring that it cannot be accidentally reconnected.
5. Allow system capacitors to discharge and check that voltage has fallen to a safe level before you start work.
6. Strip out and gain access to the high voltage battery unit.
7. Using insulated tools, disconnect the high voltage terminals from the battery unit (the high voltage terminals are often a different size from the low voltage system to try and prevent accidental cross connection).
8. Cover the exposed terminal ends in insulation tape.
9. Undo battery unit mounting bolts and remove. (Due to the weight of some battery units, you may require assistance to lift the battery from the vehicle. If assistance is used, ensure they are also wearing appropriate PPE).
10. Refit the battery unit in the reverse order of removal.
11. Reconnect power.

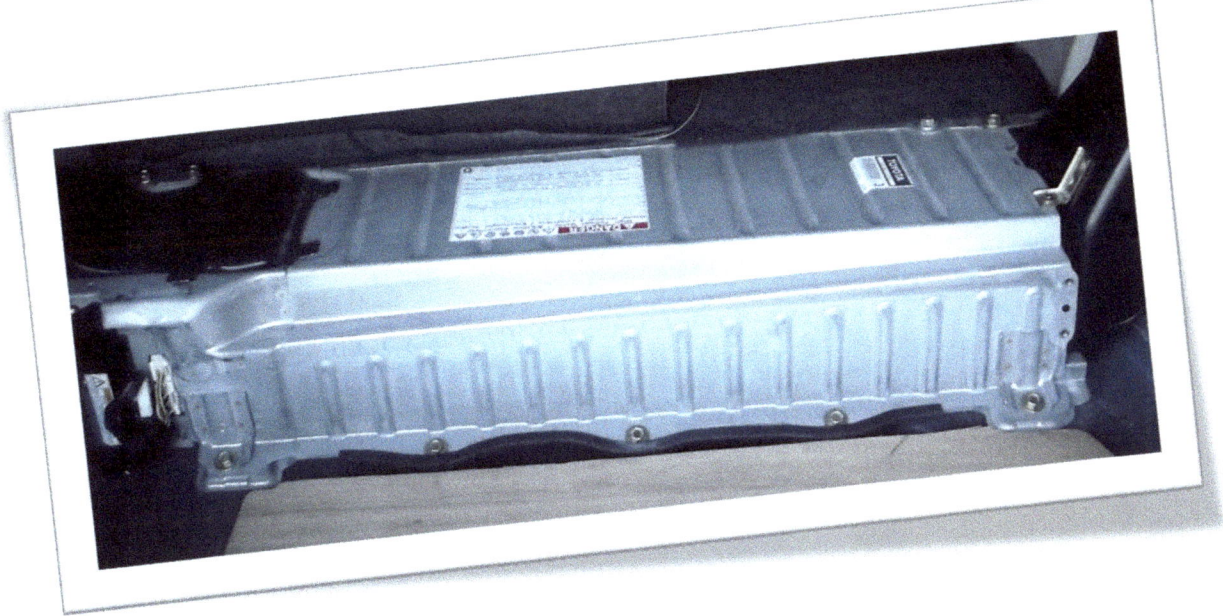

Figure 2.18 A high voltage battery unit

High voltage system capacitors

When the hybrid or electric drive system requires electricity from the high voltage battery, its delivery of power must be supplied gradually to ensure that a sudden surge of electricity doesn't damage system components. Some manufacturers use capacitors to smooth out the delivery of electricity to and from the high voltage battery units.

Figure 2.19 High voltage system capacitors

High voltage battery module relays

A series of relays are often used to control the flow of electricity from the high voltage battery. These relays will connect and disconnect both the positive and negative high voltage electrical circuits. In addition to the main connection, a further relay and resistor unit are used when initially connecting power to the system. This introduces a controlled amount of power to the system before the main relay takes over, stepping up to full system voltage.

- List the three main types of hybrid drive?
- What is meant by the term 'regenerative braking'?
- What is the ideal state of charge for a Nickel-Metal Hydride battery (Ni-MH)?

The construction and function of hybrid and electric vehicle drive motors

Two main types of electric motor are in common use on hybrid and electric drive vehicles Direct Current motors (DC ⎓) and Alternating Current motors (AC ∿).

Direct current motors

A simple direct current (DC ⎓) motor can be made by passing an electric current through a coiled wire that is wound around a central shaft called the **armature** – this creates an electromagnet.

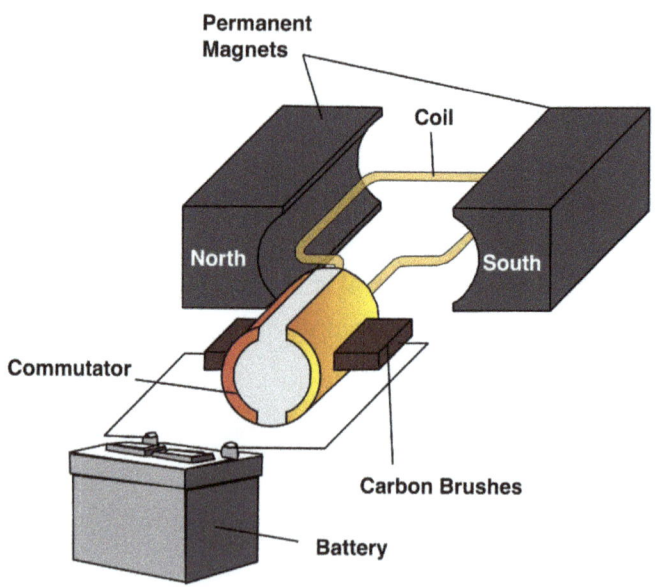

Figure 2.20 A simple electric motor

- The electric current produces an invisible magnetic field, which is repelled (pushed away) or attracted (pulled towards) by the permanent magnets surrounding it. This causes the armature to turn.

- Once the armature has turned out of the magnetic field, it would normally stop.

- To keep it rotating, the polarity of the electricity passing through the electromagnet mounted on the armature must be changed. This is done by a component called a **commutator**.

- Two spring-loaded electrical contacts called **brushes** are mounted on the end of the armature to maintain electrical connection with the commutator as the shaft rotates.

- When electric current is switched off, the motor will stop.

The stronger the magnetic forces inside the motor, the more power it will produce. The magnets inside the motor casing that surround the armature can produce stronger magnetic fields if they are wrapped in wire coils and electric current is passed through them. These external magnet wires are called **field coils.**

Armature – the central shaft of an electric motor.

Commutator – a segmented electrical contact mounted on the end of a motor armature, designed to change electrical polarity as the motor turns.

Brushes – spring-loaded electrical contacts that transfer current to the rotating armature.

Field coil – copper wiring wrapped around magnets – this can increase the magnetic field produced when supplied with electricity.

Alternating current motors AC

Another type of electric motor that can be made using magnetic fields is the alternating current (AC ∼) motor. Unlike a direct current (DC ⎓) motor where polarity has to be continuously changed through the use of a commutator, alternating current through its very nature changes direction as it operates and this can be used to swap polarity and keep the motor rotating.

AC ~ motors can be split up into different designs including:

- Brush type
- Synchronous type
- Induction type
- Three phase type

Brush type AC motors

A simple brush type AC ~ motor design uses two permanent magnets placed above and below a rotating wire coil in a similar manner to that described for DC === motors. The magnets are lined up so that the coil faces the north pole of one magnet and the south pole of the other. Conducting brushes touch two slip ring connectors, which feed the inner coil of the armature with electric current. When fed with current, magnetic forces make the armature coil rotate so that the south pole turns to the north pole of the permanent magnet, and vice versa. When the supply current alternates (changes direction), the magnetic south and north of the coil swap places and the motor continues to turn. The frequency of the alternating current will help determine the motor speed.

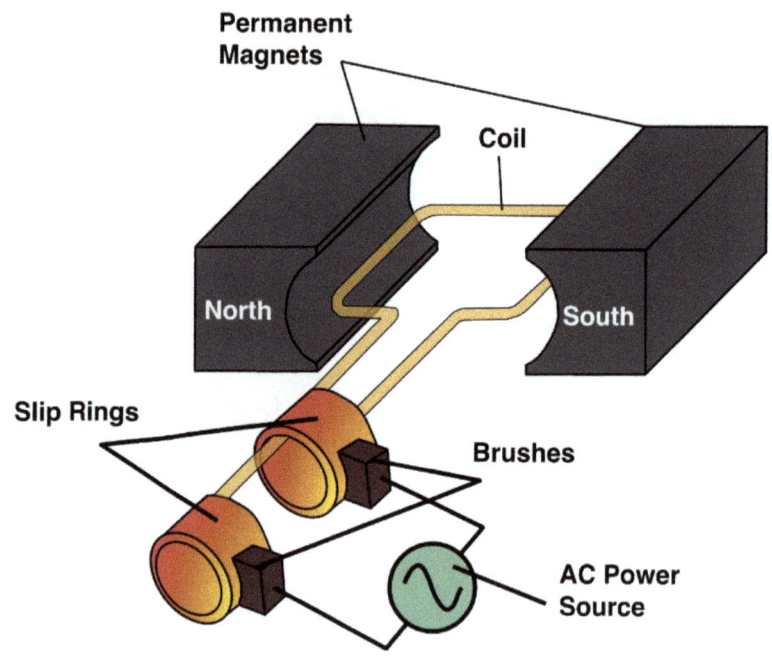

Figure 2.21 An AC brush type motor

Synchronous type AC motors

A synchronous type AC ~ motor produces a very accurate motor speed. This design has a set of coils surrounding a rotor, but instead of a wire coil for the **rotor**, it is a permanent magnet. The electric coils are arranged in opposing pairs in a stationary housing around the edge of the motor known as the **stator**. The north-south pairs attract the north-south poles of the permanent magnet on the rotor, turning it. As the alternating current cycles backwards and forwards, the motor will rotate with a very accurate speed depending on the current frequency and the number of coils used.

This type of system is known to create large quantities of heat when used in the design of hybrid electric vehicles which often requires it to have its own dedicated cooling system.

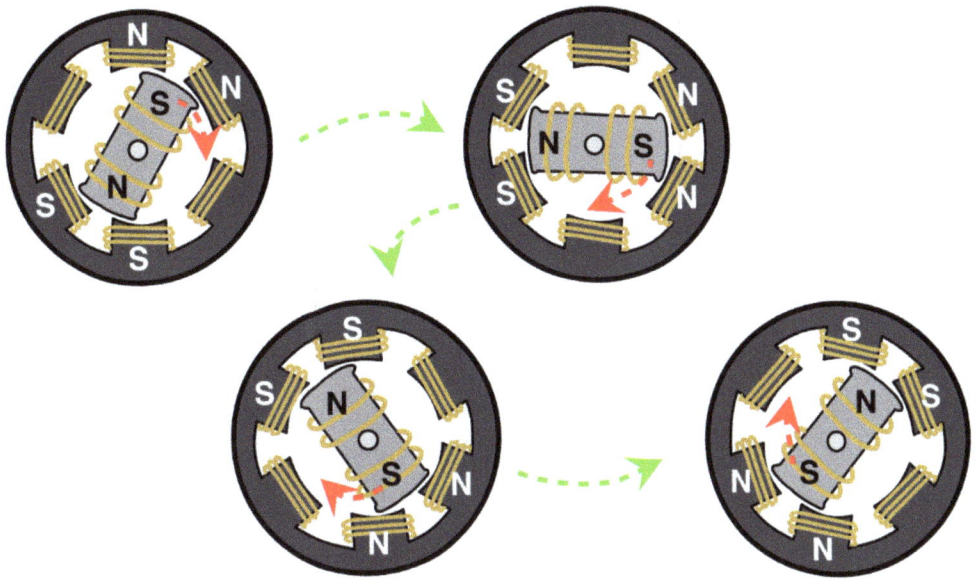

Figure 2.22 An AC synchronous type motor

Stator - the stationary, non-moving part of an electric motor.

Rotor - the turning, 'rotating' part of an electric motor.

Induction - the effect of a moving magnetic field producing an electric current in a coil of wire.

The permanent magnets used in hybrid electric motors are very strong "rare earth" magnets. Despite their name, rare earth magnets can be made from fairly common compounds and are often in the form of neodymium iron boron (NIB) or samarium cobalt. These rare earth magnets are around 10 to 15 times stronger than a standard iron magnet, making them ideal for use in powerful electric motors.

Rare earth magnets that are used in hybrid electric motors have an extremely strong magnetic attraction to each other or metal surfaces.

- They are strong enough to cause injuries to body parts pinched between two magnets, or a magnet and a metal surface; even causing broken bones or severed fingers.

- NIB magnets are very brittle and if allowed to get too near each other can strike together with enough force to chip and shatter the brittle material; the flying chips can cause injuries.

- The strength of the magnetic fields created by rare earth magnets is enough to disrupt electronic equipment such as heart pacemakers.

Induction type AC motors

An **induction** type AC ∼ motor doesn't use brushes or conducting slip rings. Instead, it uses an effect called induction, where a changing magnetic field produces an electric current in a similar way to an ignition coil. When an alternating current is passed through a series of conductor coils around the edge of the motor, they produce an invisible magnetic field. This magnetic field will induce a current in a winding attached to the armature of the motor, creating its own magnetic field, causing it to turn. This type of motor does not suffer with the wear or sparking produced by brush type motors.

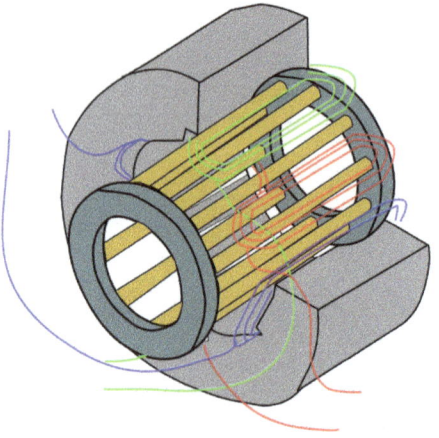

Figure 2.23 An AC induction type motor

Because the inner conductor coils on the armature of an induction type motor are shaped in a round, cage-like grid, engineers often refer to this design as a "squirrel-cage" motor.

Hybrid Electric & Alternative Automotive Propulsion

Notes: The power delivery of current hybrid electric motors is between 10Kw and 50Kw; that's around 14 to 69 horse power (UK).

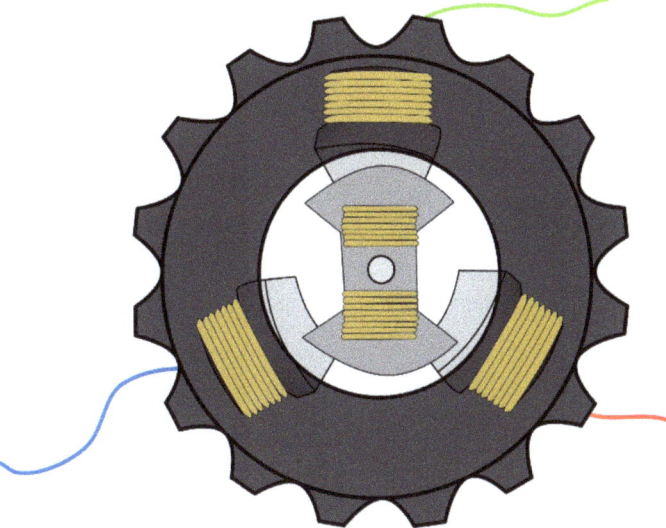

Figure 2.24 An AC three phase motor

Three phase type AC motors

Many hybrid and electric vehicles are able to use an **inverter**, which takes the high voltage DC ⎓ stored in the batteries and converts it into an AC ∿ current in three separate **phases**. Each phase is 120 degrees apart from the next, which will form a complete 360 degree cycle. Powerful hybrid electric motors are often wired for this three-phase electricity. The motors have coils spaced 120 degrees apart, each coil being driven by one phase of the electricity supply. This arrangement produces a rotating magnetic field simply and efficiently, which is able to turn a permanent magnet rotor. An advantage of this design is that it can often be air cooled, reducing weight and design costs.

Some manufacturers are beginning to produce multi-phase (more than three) electric motors, in an effort to increase the efficiency and output of their designs.

Terms:

Inverter - an electronic component that turns direct current DC into alternating current AC.

Phase - the relationship between each cycle of different alternating currents.

Hybrid drive battery charging

Not all hybrid vehicles use an alternator to charge the battery systems. In many models the electric motors used to drive the road wheels can be converted into generators. They operate when power is being provided by the internal-combustion engine or regenerative braking is occurring. These motors are able to generate the electricity required to recharge the high voltage system on a hybrid car (somewhere between 100V and 300V). The high voltage batteries are then able to charge the 12 volt battery system through a transformer called a DC to DC converter. This system reduces the amount of strain on the standard 12 volt circuits, which means that smaller batteries with a longer service life can be used for the low voltage system.

The state of charge (SOC) of the high voltage battery is constantly monitored by the charging ECU. If the charge of the high voltage battery falls below a set limit, it is possible that the internal combustion engine might start automatically to generate electricity for recharging.

The automatic starting of the engine poses health and safety risks as body parts could be caught up in moving engine components etc. It could also create engine damage if it is being worked on at the time. For example, as the engine was being serviced and the oil was drained, if the engine started unexpectedly, friction, wear and seizure could occur.

It is important that when working on, or servicing a hybrid electric vehicle, that the engine is prevented from staring automatically. This can often be achieved by storing the smart key beyond its normal range of operation, but always follow manufacturer's instructions and recommendations.

Hybrid drive starting

Although some hybrid vehicles are fitted with a starter motor, they are often only there for emergency purposes. If the high voltage battery system becomes discharged, the 12 volt starter motor cuts in to get the engine running. Once the internal combustion engine is running, electricity can be generated to operate the charging system.

During normal operation the starter motor is no longer needed, because hybrid vehicles use a combination electric motor/generator as part of their drive operation.

When the internal combustion engine needs to be started to provide drive, the hybrid electric motor can be used instead of a starter motor. The instant and powerful rotation of the crankshaft created by this motor ensures that 'stop–start' operation of the internal combustion engine is smooth and precise. As the hybrid electric motor doesn't use a starter pinion and ring gear, the characteristic whine created by a normal starter motor is no longer apparent.

The construction and function of associated hybrid components

Low voltage cabling

In the low voltage system, insulated copper wiring is used to transport electricity around the vehicle to where it is needed. Thin strands of copper (which is a conductor) are bundled together and coated with a plastic shield (which is an insulator) to help prevent electricity conducting to any other metal components. If this happened, it would cause a short circuit.

Because a large number of wires are used in motor vehicle construction, the external plastic coating is usually colour-coded (see Figure 2.25). When diagnosing an electrical circuit fault you can use these colours to help trace cable routing or identify them on a wiring diagram.

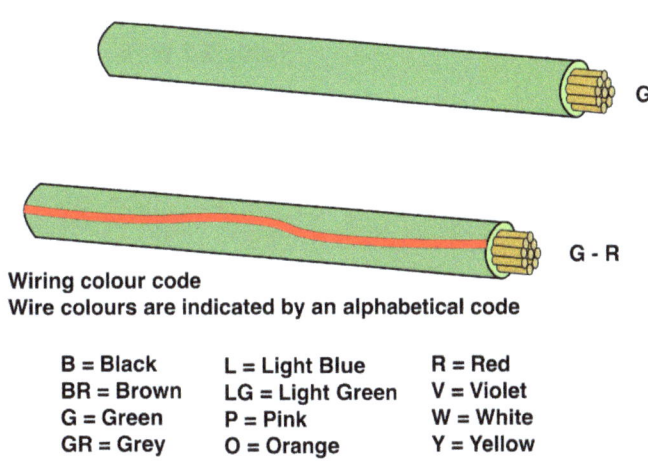

Wiring colour code
Wire colours are indicated by an alphabetical code

B = Black   L = Light Blue   R = Red
BR = Brown   LG = Light Green   V = Violet
G = Green   P = Pink   W = White
GR = Grey   O = Orange   Y = Yellow

The first letter indicates the basic wire colour and the second letter indicates the colour of the stripe.

Figure 2.25 Low voltage electric wire with common electrical colour codes

Electrical wires come in different sizes. Copper strands are bundled together, which means that if one or more strands are damaged electricity can still flow. Automotive low voltage wires are normally labelled with the number of strands they contain and the diameter of each strand in millimetres. This gives an indication of the amount of current the wire is able to carry.

Some typical wire size designations and uses are shown in Table 2.7.

The thicker the wire, the more electricity it can carry and the less internal resistance it has. This means that more voltage and current will be available for the component.

The longer the wire, the higher the resistance. This means that less voltage and current will be available by the time it reaches the component.

- Double the length of the wire and you double the resistance.
- Double the diameter and you halve the resistance.

Table 2.7 Wire sizes and their uses

Number of strands / Wire diameter	Continuous current rating	Uses of the wire
9 / 0.30mm	5.75 amps	Side lamps, tail lamps, reversing lamps, horns
14 / 0.30mm	8.75 amps	Side lamps, tail lamps, reversing lamps, horns, general wiring
28 / 0.30mm	17.5 amps	Headlamps, fog/driving lamps, windscreen wiper motor
44 / 0.30mm	27.5 amps	Charging cable (after DC to DC converter), battery feed
65 / 0.30mm	35 amps	Charging cable (after DC to DC converter)
84 / 0.30mm	42 amps	Charging cable (after DC to DC converter)
97 / 0.30mm	50 amps	Heavy duty charging cable (after DC to DC converter)
120 / 0.30mm	60 amps	Heavy duty charging cable (after DC to DC converter)
80 / 0.40mm	70 amps	Heavy duty charging cable (after DC to DC converter)
37 / 0.71mm	105 amps	Emergency starter/Battery cable
37 / 0.90mm	170 amps	Emergency starter/Battery cable
61 / 0.90mm	300 amps	Emergency starter/Battery cable

Terminals, connectors and continuity

When a manufacturer designs a car, the electrical wiring used to create the circuits can be bundled together as insulated sections called wiring looms. When the car is assembled, these looms can be routed around the vehicle in the most efficient way, and hidden from view behind panel work, carpets and trims. The wiring looms are made in sections and joined together by **connectors**. At the ends of the looms, **terminals** are used to connect the wires to electrical components.

For electricity to operate the components correctly the circuits must be continuous and unbroken – this is called **electrical continuity**.

The terminals of the high voltage electric circuits are often different sizes from the low voltage system to prevent accidental, incorrect connection.

Low voltage earth return systems

Vehicle designers and manufacturers try to keep the amount of wiring used to a minimum. This will save on materials, improve efficiency and reduce costs. Because many vehicles are manufactured mainly from metals (which are good conductors of electricity), it is not always necessary to complete an electrical circuit back to the battery using wire alone.

- The negative end of an electrical circuit wiring can be connected to the vehicle body or chassis. This is called an earthing point.

- The negative terminal of the battery can also be connected to the vehicle body or chassis to complete the circuit. This is called **earth return**.

Connector – a component that joins two parts of a circuit together.

Terminal – where the circuit ends (terminates).

Electrical continuity – when an electrical circuit conducts current easily and is unbroken (i.e. continuous), it has electrical continuity.

> Because the metal of the vehicle body now forms part of the car's electrical system, it is possible to cause damage or injury by mistakenly allowing the positive side of the electrical circuit to come into contact with this **earth** return. The most common time for this to happen is when you are replacing the hybrid car's low voltage battery.
>
> To reduce the possibility of injury or damage, you must follow the procedure described below when connecting or disconnecting a car battery.

Disconnecting and connecting the battery

When you connect or disconnect the low voltage battery from a vehicle you should remove and refit the terminals in a certain order, as this will help reduce the possibility of a short circuit.

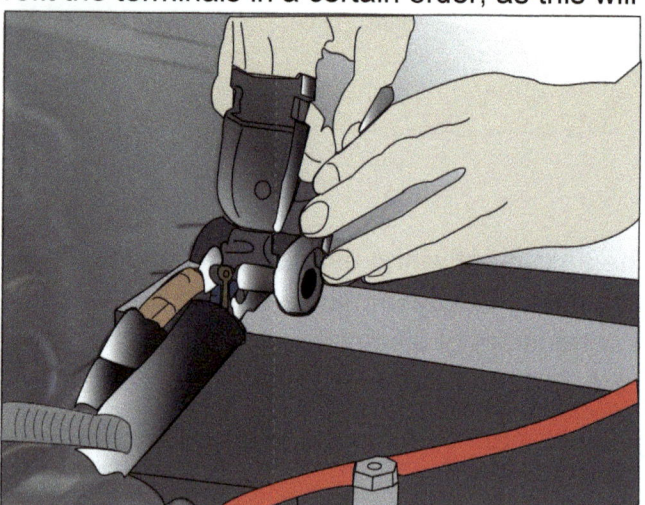

Figure 2.26 A battery terminal being disconnected

Order for disconnecting: Always remove the negative lead first when you are disconnecting the battery. Once you have disconnected the negative terminal, the vehicle's electrical system is now **open circuit**. Therefore, if the spanner you are using accidentally touches the car's bodywork, they both have the same electrical potential or pressure. If the pressure in an electric circuit is equal, no current can flow (because a difference in pressure creates flow).

Order for connecting: When reconnecting a battery, always connect the positive terminal first and the negative terminal last, for the same reasons.

During the connection and disconnection of electrical components on hybrid and electric vehicles, it is possible to cause extreme damage to the vehicle and personal injury. If components or systems are left switched on, it is possible that excessive current draw may be converted into heat. Always ensure that no power will be discharged through an electric circuit or component during the connection of an additional 12 volt power source.

Earth – an electrical ground connection, designed to complete a circuit.

Earth return – using the metal chassis to complete the car's electrical circuit.

Open circuit – a broken electrical circuit where no electricity can flow.

High voltage cabling

In the high voltage system, different demands are placed on the wiring, leading to special designs and construction. The main factors that must be considered in the design of high voltage cabling used in hybrid and electric vehicles are:

- Large power transmission between the high voltage battery and the engine motor/generator
- Shielding against the electromagnetic interference caused by the power electronics
- Large differences in temperature due to ambient conditions and electrical loading
- Safety, armouring and colour coding of the external insulation
- Flexibility due to tight installation space

The transmission of power between the battery and motor/generator unit is achieved by using large diameter cabling, which is able to handle the amount of current required by the system. It can be single or multi-cored and made from copper or aluminium. The design of the core conductor will have a direct effect on the cables flexibility for use in tight installation spaces.

The reduction of electromagnetic interference is accomplished by using a shielding made from braided copper strands that surround the conductor, underneath the insulation.

The materials used in the construction of the high voltage wiring will allow it to operate in temperature ranges between -40º C and 200º C, which will take into account weather and heat created by high current draw during electrical operation.

The external insulation of the high voltage wiring is often reinforced to reduce the possibility of accidental cutting and is coloured bright orange as a warning that it contains high voltage electricity. The terminals of high voltage caballing are often of a different size to low voltage systems to try and prevent accidental connection to incorrect circuits which may cause damage or injury.

Figure 2.27 High voltage system cabling

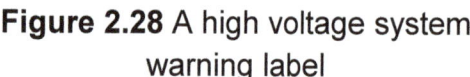

As an added safety precaution, many manufacturers provide labelling and safety information on or around high voltage systems or cabling. These labels are necessary to provide a warning and should not be removed.

Figure 2.28 A high voltage system warning label

Hybrid Electric & Alternative Automotive Propulsion

Insulated return

There are certain safety issues that must be considered when a vehicle is designed with an earth return system. In a high voltage system, if the vehicle body was used as part of the vehicle's electrical circuit, it has the potential to become live, causing electric shock and even death.

Because of this, the high voltage systems of hybrid drive vehicles have an insulated earth system. These vehicles are designed so that the wiring of the various high voltage electric circuits terminate at an electrical connector called a **bus bar** (see Figure 2.29). The bus bar is insulated from the body or chassis. A single insulated earth return wire can now connect the bus bar back to the negative side of the battery.

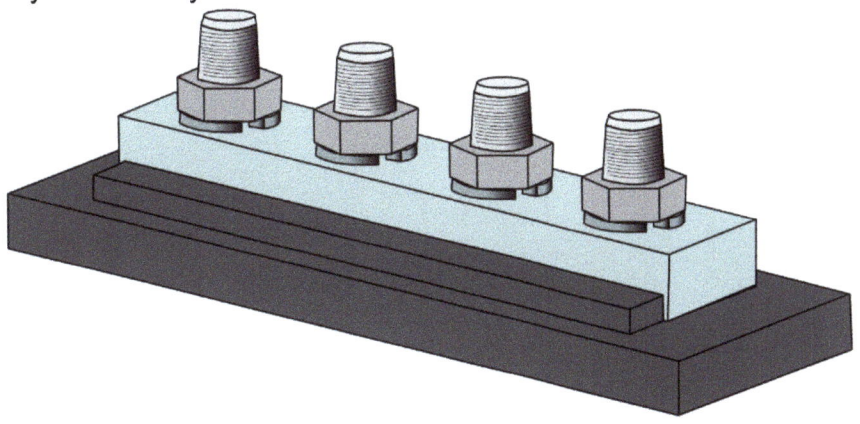

Figure 2.29 Bus bar

Bus bar – an electrical connector that makes a single connection between several circuits.

Fuse – a weak link in an electrical circuit designed to burn-out if excessive current flow occurs, preventing extra damage.

Circuit protection

Fuses and circuit breakers

If you allow electrical current to flow in a circuit without passing through a consumer, then the energy will be converted into heat. This can rapidly cause the circuit wiring and components to burn out, which can be extremely dangerous and expensive to repair.

To protect the circuit, fuses or circuit breakers are commonly used. A **fuse** is a weak link placed in the circuit. It is a thin piece of wire (with a current rating just above that of the current intended to flow in the system) that is fitted in-line (series).

This relatively inexpensive component is designed to burn out if a rapid increase in current flow occurs and prevent any further damage to the rest of the circuit. Once the fuse has burnt out, an open circuit exists and no further current can flow (electricity stops), hopefully saving more expensive parts. Fuses come in different sizes, shapes and types, including glass, ceramic and blade. Blade fuses are the most common type found on light vehicles today.

Some manufacturers are also using circuit breakers in their electrical system designs. These perform a similar function to fuses, in that they create an open circuit if excess current is allowed to flow in the electrical system. Unlike fuses however, a circuit breaker can often be reset after it has been 'tripped' meaning it is reusable.

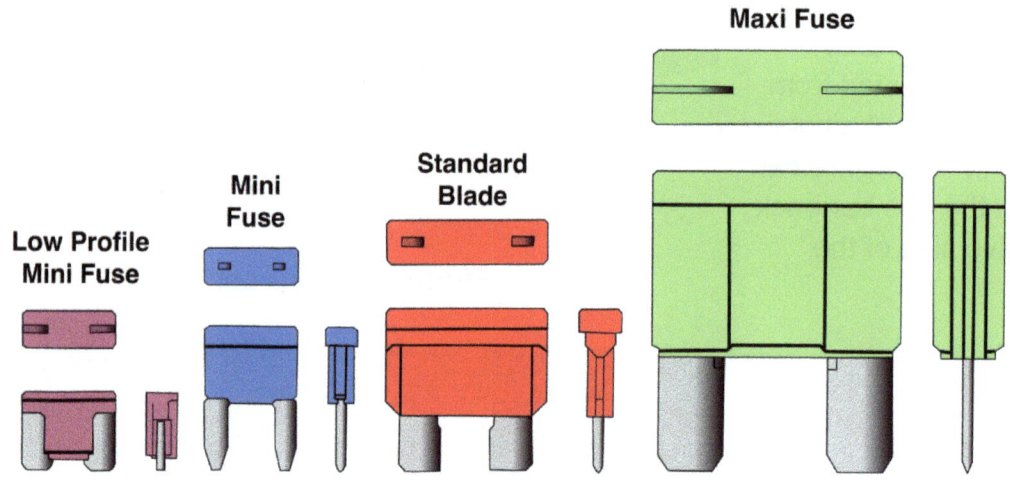

Figure 2.30 Blade fuses

Control units

Battery control module (BCM)

The battery control module monitors the operation of the high voltage battery system and helps ensure a constant state of charge SOC. Battery temperature can be measured using strategically placed thermistors. The battery temperature will give a good indication of the state of charge of a metal hydride battery because as a high charge capacity is reached, excess electrical energy will be converted into heat. The battery control module is able to activate electric cooling fans which will draw air through the battery housing and help to maintain a constant temperature.

Motor control module (MCM)

The motor control module manages the operation of the hybrid system motor/generator. When required it can operate the electric drive motor in a mode that provides full propulsion or assistance to the internal combustion engine. It will also regulate the charge when the motor operates as a generator in conjunction with regenerative braking.

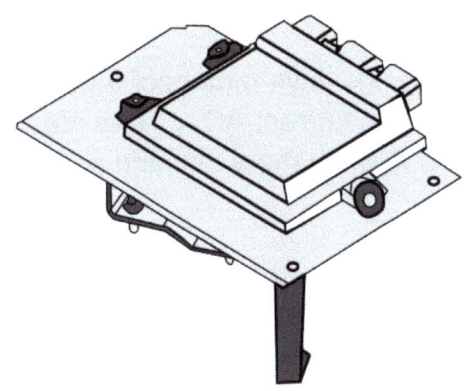

Figure 2.31 Motor control module

Distribution units

Power control unit

The power control unit manages the distribution of electricity between the high voltage battery and the motor generator unit. It will also help distribute the electricity to the DC ⎓ to DC ⎓ converter for charging the low voltage battery.

DC to DC converter

The DC ⎓ to DC ⎓ converter manages the charging of the cars low voltage battery. High voltage is taken from the metal hydride battery and stepped-down in order to charge the 12 volt lead acid battery. This means that the system no longer requires an alternator, reducing the loads placed on the internal combustion engine, increasing performance, reducing exhaust emissions and improving fuel economy. As strain on the 12 volt battery system has been lowered, and the unstable charging usually provided by an alternator has been removed, a smaller battery with a longer service life can be obtained.

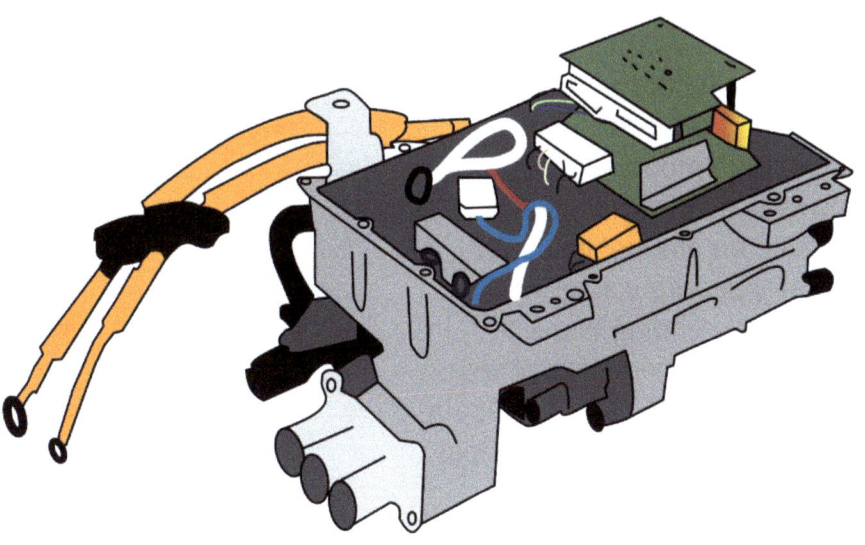

Figure 2.32 DC to DC converter

Inverter

The electric drive motors of a hybrid car will often operate using high voltage, three phase alternating current AC ∿. This means that a device is needed to take the high voltage DC ⎓ stored in the battery unit and convert it back into high voltage AC ∿; this is the job of the inverter.

The inverter will contain a transformer which is used to step-up the electric voltage from the battery into the higher voltage required by the motors. A step-up transformer works by having two coils of wire in close proximity; a primary and a secondary. When electricity flows in the primary winding, an invisible magnetic field is created which cuts across the secondary winding. To induce a voltage in the secondary winding the magnetic field has to move. This is done with a component called an 'oscillator'. The oscillator will normally contain a series of electronic switches known as insulated gate bipolar transistors (IGBT). The IGBT's are good at handling large amounts of power with very fast switching rates. They are arranged in a layout similar to the diode pack of a charging system rectifier, and as they operate will allow the primary winding to switch at great speed with varying pulse width modulation (PWM). The switching is then able to produce an alternating current with stepped up voltage in the secondary winding of the transformer. The oscillator will use several IGBT's to produce a three phase high voltage output with varying frequency. The varying frequency can then be used to provide power to the hybrid drive motor with different speed and torque capabilities depending on driving requirements.

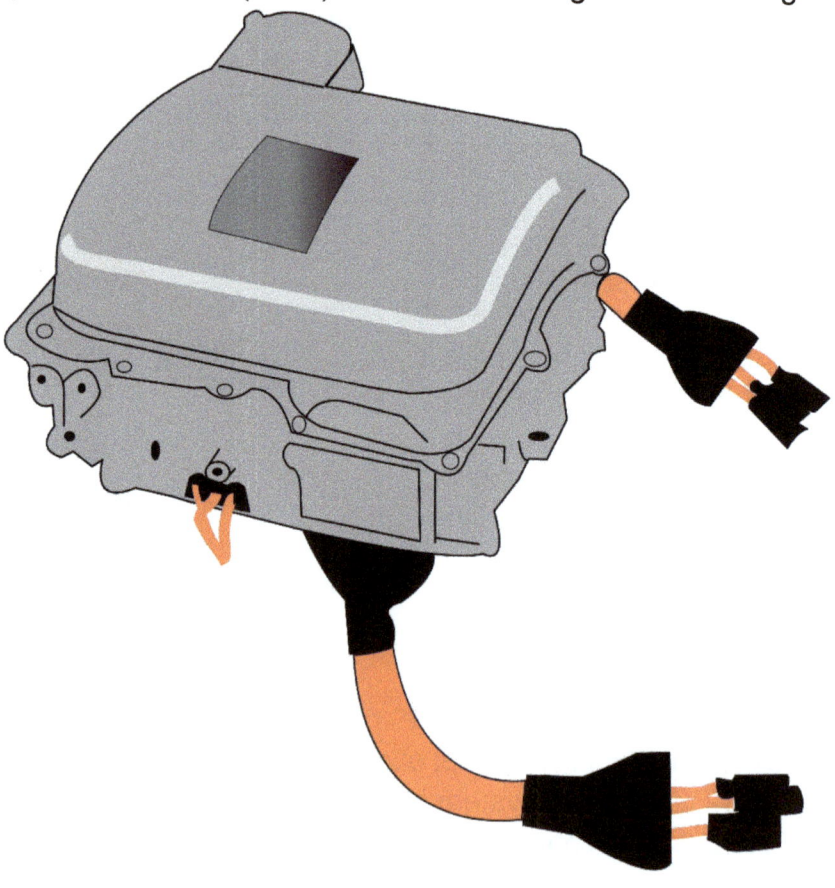

Figure 2.33 Inverter

Safety precautions to be taken before carrying out any maintenance and repair procedures on hybrid vehicles

Typical voltages used for a range of electrically propelled and hybrid vehicles are 100-650V. Working with high voltage systems found on hybrid vehicles and electric vehicles will require the use of specific personal protective equipment.

ECE R100 (relating to vehicle regulations) paragraph 2.14 clearly defines high voltage:

'High Voltage means the classification of an electric component or circuit, if its working voltage is > 60 V and ≤ 1500 V DC or > 30 V and ≤ 1000 V AC root mean square (rms).'

NOTE: This is different to definitions in commercial and domestic use which are:

Extra Low Voltage <50 V rms AC and <120 V DC
Low Voltage 50-1000 V rms AC and 120-1500 V DC
High Voltage >1000 V rms AC and >1500 V DC

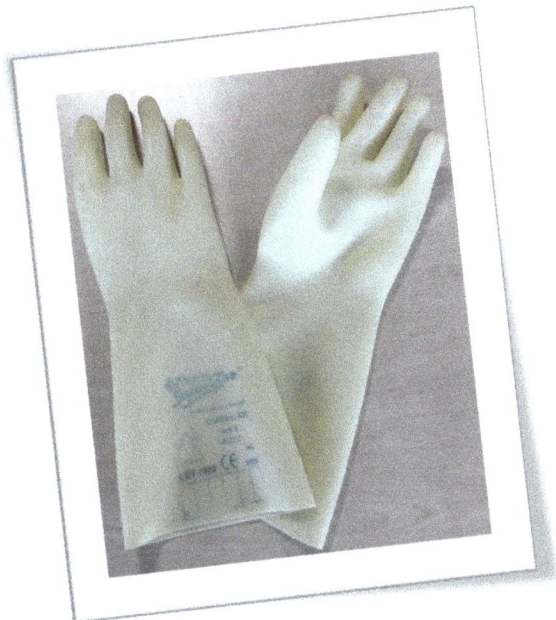

Figure 2.34 High voltage safety gloves

Table 2.8 describes the recommended PPE and reasons why it should be used.

Table 2.8 PPE required when working on high voltage systems

PPE	Recommendations
Overalls	It is recommended that overalls should have non-electrically conductive fasteners. Metal fasteners may also create issues when working around the extremely strong magnets used in hybrid system motor/generators.
Gloves	Rubber insulating gloves are one of the most important pieces of personal protective clothing for working on or around high voltage vehicle electrical systems. Gloves should be rated at least Class 0—Maximum use voltage of 1,000 volts AC, proof tested to 5,000 volts AC, and be in good condition with no rips or tears. Before use you should blow into the open wrist section and then roll the glove up from that point to see if it is able to hold the air. If any leaks are present the gloves should be replaced.
Protective footwear	Safety footwear is important when working on any vehicle system as reinforced toe caps provide protection from crush injury. When working with high voltage systems there may be occasions where insulated rubber boots might be necessary. (When a vehicle has been submerged in water for example, as this adds to the risk of electric shock). Insulating rubber boots or insulating rubber over-boots are available which will help provide extra protection.
Goggles	As with any electrical system, sparks created by accidental discharge through a short circuit have the potential to cause damage to the eyes. Safety glasses will help provide some defence from this, but goggles and face shields that fully enclose the eyes will give the most protection.

Other safety precautions that should be observed when working with vehicle high voltage systems are shown in Table 2.9.

Table 2.9 Additional precautions

Hazard	Precautions
Using electrical equipment with high voltage systems	Wear appropriate PPE. Where possible, use electrical hand tools that are specifically designed for use with high voltage systems. When using a multimeter, ensure that your fingers are behind the insulating finger guards of the test probes. If the multimeter screen shows a low battery indicator, replace the battery immediately. A low battery in the multimeter, may lead to incorrect readings being taken which could lead to injury or death.
Disposal of waste materials	Under the Environmental Protection Act 1990 (EPA), you must treat old batteries (lead acid and metal hydride) as hazardous waste and dispose of them in the correct manner. They should be safely stored in a clearly marked container until they are collected by a licensed recycling company. This company should give you a waste transfer note as the receipt of collection.
Dealing with leakage	Wear appropriate PPE. If battery leakage occurs, cover the spill with a neutralising agent such as sodium carbonate or baking soda mixed with equal parts water. After neutralising, rinse the contaminated area with clean water. If the spill involves a large amount of electrolyte, call the fire services and allow them to handle it. This may help prevent you getting seriously hurt and reduce any environmental issues.

Table 2.9 Additional precautions

High voltage electrical system isolation	When working on or around the vehicles high voltage components, you should correctly isolate and insulate the system, this should involve:
	Wear appropriate PPE.
	Make others aware that the high voltage system is being worked on.
	Remove the ignition key (if the ignition key cannot be removed, due to damage for example, take out all of the fuses in the fuse box).
	If the ignition key is a 'smart key' which doesn't require insertion into a key slot, it should be safely stored more than thirty feet away from the vehicle.
	Disconnect the low voltage (12 volt) auxiliary battery.
	Following manufacturer's instructions, remove the high voltage service plug or switch and lock the high voltage isolator button so that they cannot be accidentally reconnected.
	Allow time for any system capacitors to discharge and check the voltage has fallen to a safe level using a multimeter.
	Do not cut any orange high voltage cable.
	Cover any disconnected terminals with insulating tape.
Submerged vehicle safety	To handle a hybrid or electric vehicle that has been partially or fully submerged in water the high voltage system and air bags will need to be safely isolated.
	Wear fully insulating electrical PPE (as described in table 2.8).
	Immobilise vehicle and remove ignition key (if the key is a 'smart key', ensure it is kept beyond its range of operation).
	Where possible allow any water to drain/dry.
	Remove all system fuses and isolate the high voltage system as described above.

Table 2.9 Additional precautions

Highly magnetic components	The magnets used in hybrid electric motors are around 10 to 15 times stronger than a standard iron magnet. The naturally high magnetic field produced can affect the correct operation of heart pacemakers.
	Care should be taken to remove all metal jewellery and avoid the use of delicate electronic equipment such as mobile phones while working on these systems.
	If the rotor of a hybrid drive motor needs to be removed/fitted, it will be necessary to use a special tool, so that the magnetic attraction to other metal components does not affect its fitment and create damage to the motor or cause personal injury.
Medical conditions that may be affected by high voltage or magnetic fields	Existing medical conditions such as heart conditions can be affected by both the very strong magnetic fields produced in hybrid drive motors and the high voltage systems.
	It is not recommended that people with heart pacemakers work on these systems.
Checking voltage prior to working near or on high voltage systems	Before any work is started on a high voltage system, the electrical circuits should be isolated and capacitors allowed to discharge. The system voltage should be checked with a multimeter to see that it has fallen low enough to begin any work.
	Voltages higher than 60 volts DC or 30 volts AC RMS are likely to cause electric shock leading to injury or even death.

Following any repairs that have involved the disconnection of either the high or low voltage battery systems, it is important to correctly reinstate the condition of the vehicles electrical systems and ensure their correct function and operation. This includes:

- Using a scan tool to ensure that no fault codes have been created
- Resetting any on-board displays
- Re-programming electric window one touch operation

Reinstate high voltage

How to:

1. Ensure that you are wearing the correct PPE.
2. Make sure that all electrical consumers are switched off.
3. Remove the key from the ignition.
4. If the ignition key is a 'smart key', make sure that it is stored away from the car, outside its range of operation.
5. Using insulated hand tools, reconnect the high voltage system terminals, being careful to observe the correct polarity and avoid short circuits.
6. Reset high voltage system isolator switch or service plug to the on position.
7. Insert ignition key and operate all electrical systems to check for correct function and operation.
8. Reset any clocks, dashboard displays and one touch electric window programming.
9. Use a scan tool to ensure that no diagnostic trouble codes have been generated by the connection of the high voltage system.

Figure 2.35 A high voltage isolation service plug

- List the four main designs of AC electric motor.
- What component is used to charge the low voltage battery of a hybrid drive vehicle?
- In a hybrid or electric drive vehicle which component turns direct current DC from the battery to alternating current AC for the motors?

Internal combustion engines

In order to be considered a hybrid, the electric motor drive system is coupled to an internal combustion engine. The vehicle drive can then be shared between the systems as series, parallel or a combination of the two (see 'system layouts').

The operation of the internal combustion engine relies on the four stroke cycle, which includes:

- Induction - air and fuel are introduced to the engine cylinder
- Compression - the air and fuel mixture is compressed to a much smaller volume in the combustion chamber
- Power - the air fuel mixture is ignited and the pressure created by the heat of the burning mixture is used to drive pistons, which in turn rotate the crankshaft
- Exhaust - the waste products created during the combustion process in the engine are expelled to atmosphere

For more information on the operation of internal combustion engines, see Chapter 3.

Most hybrid vehicles combine an electric motor with a spark ignition petrol engine. This is because it provides the most efficient use of torque distribution between the two motor drive systems (internal combustion engine and electric motor).

A spark ignition petrol engine produces its most efficient delivery of torque high up in its rev range. When this is combined with the instantaneous torque delivery from an electric motor, it provides good all round capabilities.

A compression ignition Diesel engine produces its most efficient delivery of torque lower down in its rev range. When combined with the instantaneous torque delivery from an electric motor, it is often not as effective as a petrol engine combination and therefore a less popular design.

With the addition of an effective transmission system, some manufactures are using Diesel engines with their hybrid drive to achieve very high fuel efficiency.

To improve the efficiency of spark ignition engines used in combination with hybrid drive, manufacturers have developed engines that are able to adapt their operating cycles depending on load and speed requirements. Two examples of these designs are described in the next section.

Atkinson cycle

A true Atkinson cycle engine is able to vary the length of its stroke depending on its position in the four stroke cycle. The piston and connecting rod are linked to the crankshaft using a pivoting intermediate arm as shown in figure 2.36. As the crankshaft rotates through one complete revolution, the intermediate arm will provide a different angle of connection to the big end of the con rod during 180 degrees of rotation. This can give the engine a different length induction and compression stroke from its power and exhaust stroke. By reducing the length of the induction and compression stroke and maintaining the length of the power and exhaust stoke, the engine is able to make effective use of the thermal energy contained in the air/fuel mixture at the expense of power. This engine type produces very good fuel economy but would not be suitable for use in driving situations demanding a high power output.

Figure 2.36 An Atkinson cycle engine

Modern engine designers have come up with a method of creating an internal combustion engine that uses the principles of an Atkinson cycle without compromising performance when driving situations dictate the need for power. A similar effect to changing the length of stroke during engine operation can be achieved by variable valve control. If during the induction and compression stroke, the inlet valve is held open for longer than the normal time period, some of the air drawn in during the induction stroke can be pumped back into the inlet manifold during the compression stroke. This reduces the volume of air compressed, giving similar results to changing the length of the engines induction and compression stroke.

Fuel can then be injected directly into the combustion chamber and ignited by the spark plug to create the power stroke. The power stroke uses up all of the heat density in the burning air/fuel mixture improving the efficiency, but giving a reduced power output. During the exhaust stroke, the exhaust valve is operated for the standard period of time allowing all of the exhaust gasses to escape in the normal manner.

The advantage of a system that simulates the Atkinson cycle through variable valve control is that it can be altered to run in a conventional manner when power is needed, providing alternative operating methods for efficiency, economy, emissions and performance.

Notes: To further reduce emissions and improve fuel economy on their internal combustion engines, some manufacturers use an insulated storage reservoir of engine coolant. The reservoir keeps a quantity of hot coolant for around three days and when the vehicle is first started an electric pump is able to circulate it around the engine providing a pre-heating system. This ensures that the engine reaches its optimum operating temperature in a very short period of time, improving its overall efficiency.

Valve lift control

VTEC

The VTEC system is a method used to control how far the inlet valves open. Different cam profiles are machined on the same camshaft, and rocker arms are used to transfer the movement from the different cam profiles to the inlet valves when required. This design has been used for some time to gain extra performance and volumetric efficiency from engines and its operation is described in the next section.

A shallow cam profile will only open the inlet valve a short distance, whereas an aggressive cam profile will open the valve fully. At low engine speed and load, a shallow cam profile is used to provide smooth running, fuel economy and lower emissions. At high engine speed and load, an aggressive cam profile can be used to provide performance.

To switch between the different cam profiles, hydraulic oil pressure is used to move a locking pin between two rocker arms. At slow engine speed, the shallow cam profile operates the low lift rocker arm to open the valve; the high lift rocker arm moves freely against a return spring (idles) with no effect on engine operation.

As engine speed increases, hydraulic oil pressure locks the high lift rocker arm to the low lift rocker arm. The aggressive cam profile now takes over, opening the inlet valve fully. As engine speed falls, hydraulic oil pressure is directed away from the locking pin and a return spring is used to unlock the two rocker arms and the low lift cam takes over. Due to the nature and operation of the VTEC system, the change in performance can often be felt by the driver as they accelerate.

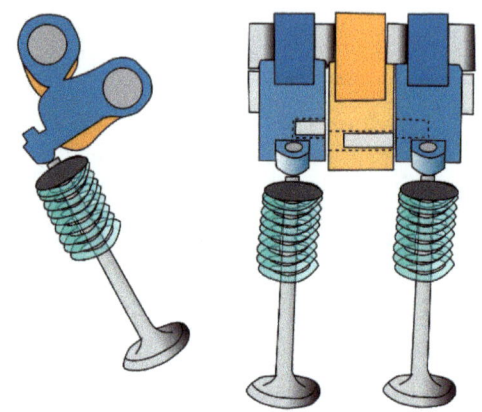

Figure 2.37 VTEC control

In addition to this performance enhancing valve lift system, a 3 stage operation can be incorporated which also stalls the action of the engines valves on overrun, sealing the cylinders, stopping combustion and reducing the normal pumping losses caused by induction and compression. This reduces overall **engine braking** and allows more electricity to be created during regenerative braking.

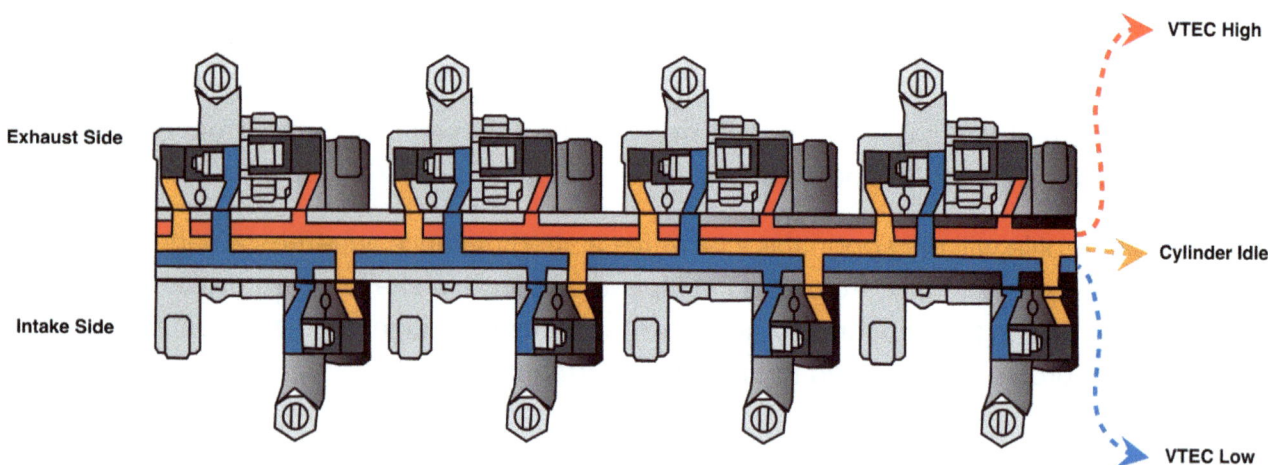

Figure 2.38 Three stage VTEC operation

Aerodynamics and body shape

Even with the low emission and fuel saving combinations of electric motors and specialist internal combustion engine design, efficiency can be improved even more with **aerodynamic** body styling.

Any vehicle moving through air will encounter resistance caused by **drag** which often result from body shape/design and friction created by the material of the body shell. As a car is driven forwards, two forces oppose the movement. A high air pressure is created at the front of the car and a low air pressure is created at the rear. If the body is aerodynamically designed, using shapes that reduce these two opposing pressures, overall efficiency is improved, this is why many hybrid and electric vehicles may have unusual **aesthetics**.

Certain design features of cars will also create drag as the air moves over the surface of the vehicle while in motion. By integrating body features such as bumpers and mirrors, and enclosing wheels inside bodywork, drag can be reduced further still.

Engine braking - the slowing of the vehicle, created by resistance from the engine during overrun.

Aerodynamic - having a shape that reduces drag as air moves past.

Drag - a retarding force which tries to slow something down.

Aesthetics - how nice something looks.

Transmission

A hybrid vehicle using a combination of a petrol engine and an electric motor produces a combination of torque ranges.

- An electric motor gives its greatest amount of torque when starting.
- A petrol engine gives its greatest amount of torque at speed.

Many manufacturers use a continuously variable transmission (CVT) with hybrid vehicles which will deliver the most efficient amount of torque and speed to the road wheels no matter which motor is driving at the time.

Continuously variable transmission (CVT)

A continuously variable transmission (CVT) gearbox is a form of automatic transmission. Although the design has been around for over 100 years, many manufacturers are now offering this type of gearbox as an option because of its efficient delivery of torque and power. Instead of using a fixed set of five or six gear ratios, this type of transmission is able to provide a step-less ratio between an upper and lower limit. This means that when coupled to an engine, it is always able to run within its optimum range.

Two main types of CVT are used:

- variable diameter pulley (VDP)
- toroidal CVT

Variable diameter pulley (VDP)

Instead of using mechanical gear sets, as in an **epicycle**, VDP uses a drive belt held between two pulleys similar to the chain and sprockets on a bicycle. Originally this belt was made of rubber but, as technology has developed, a steel drive belt has replaced the earlier design.

A bicycle is able to vary its gearing by changing the size of the sprocket on which the drive chain runs. VDP operates by changing the size of the drive pulleys, which allows different gear ratios to be created. To do this the drive pulleys expand and contract. In this way, the drive belt is able to ride up and down within the pulleys, varying their size and therefore the gear ratio. As one pulley expands the other will contract equally.

The steel drive belt is made up of many small links held together on a metal band. As the drive pulleys rotate, the metal links are forced into compression, causing the belt to push rather than pull.

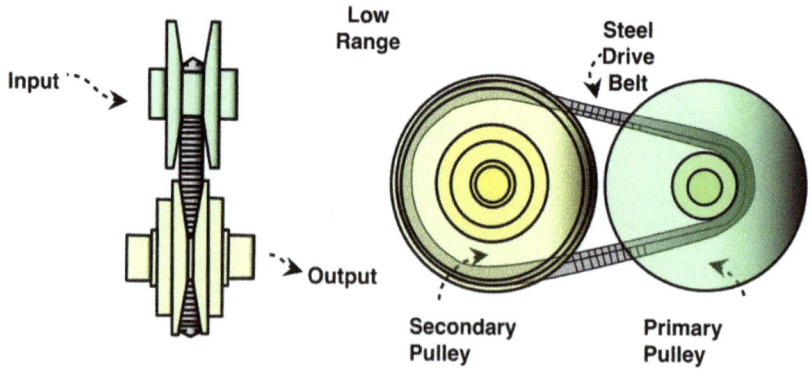

Figure 2.39 CVT low gear range

Because these pulleys do not rely on fixed gear sizes, a stepless gear ratio can be achieved, which maintains optimum efficiency for any engine speed or load. The output from the drive pulleys is normally transmitted through a further reduction gear, which can be of epicyclic design. This also allows a reverse gear to also be included.

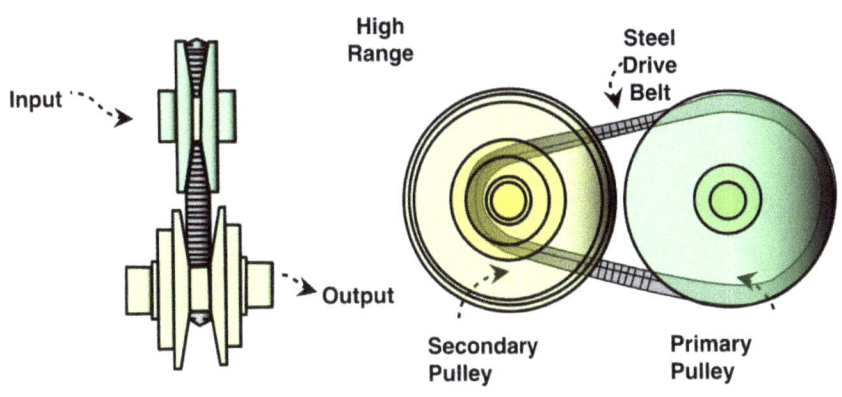

Figure 2.40 CVT high gear range

Toroidal CVT

A toroidal CVT has a tapered input disc and output disc which are placed face-to-face to form a **concave** driving surface. The input and output discs are able to turn independently of one another and are connected using **torus**-shaped rollers. The rollers are able to ride up and down against the concave surface of the input and output drive discs and transfer turning effort between the two.

- When the roller is touching a low point on the input disc curve and a high point on the output disc curve, a low gear ratio is achieved.

- When the roller is touching a high point on the input disc curve and a low point on the output disc curve, a high gear ratio is achieved. By moving the rollers across the surfaces of the input and output discs, a continuously variable transmission (CVT) ratio can be achieved. The output from the drive disc is normally transmitted through a further reduction gear, which can be of epicyclic design. This allows a reverse gear also to be included.

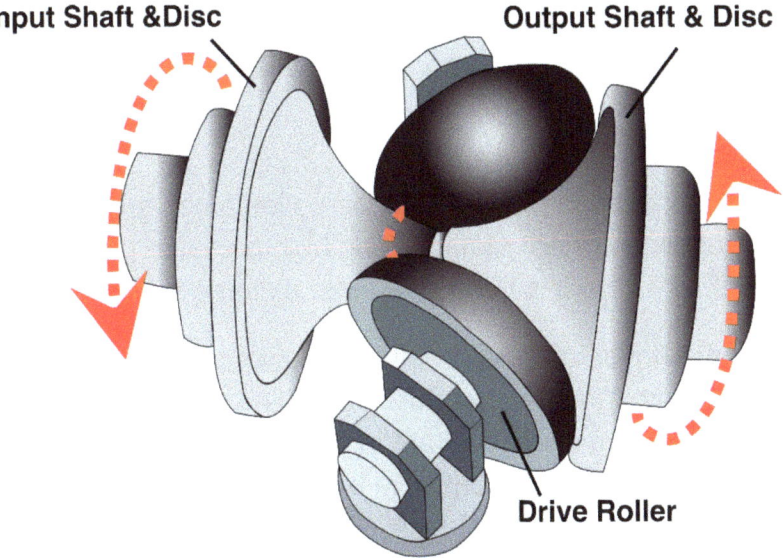

Figure 2.41 Toroidal CVT

Epicycle - a type of planetary gear set used in some automatic transmissions.

Concave – curved inwards.

Torus – ring-shaped like a doughnut.

Air-conditioning and climate control on hybrid and electric vehicles

Conventional systems

A conventional air-conditioning or climate control system uses an engine-driven pump called a compressor to raise the pressure of a refrigerant gas in a sealed system. The most common gas currently used is Tetrafluroethane, known as R134a. Other refrigerant gases include:

- Tetrafluoropropane - HFO1234 yf
- Dichlorodifluoromethane – R12 (now obsolete)
- Carbon dioxide – R744.

The gas passes through a radiator, called a condenser, which is normally mounted just in front of the cooling system radiator. The high pressure gas is then cooled and condensed into a liquid. From here it is transferred into a storage unit called a receiver drier until it is needed.

When the driver operates controls to lower the cabin temperature of the car, the refrigerant is released through a temperature-controlled expansion valve (TXV). As the pressure falls, the liquid refrigerant changes state in another small radiator inside the car called an evaporator. The temperature in the evaporator falls, and as the cabin air is circulated through it, heat is removed. This helps cool the air inside the car. The refrigerant is then returned to the compressor, where the whole process starts once again.

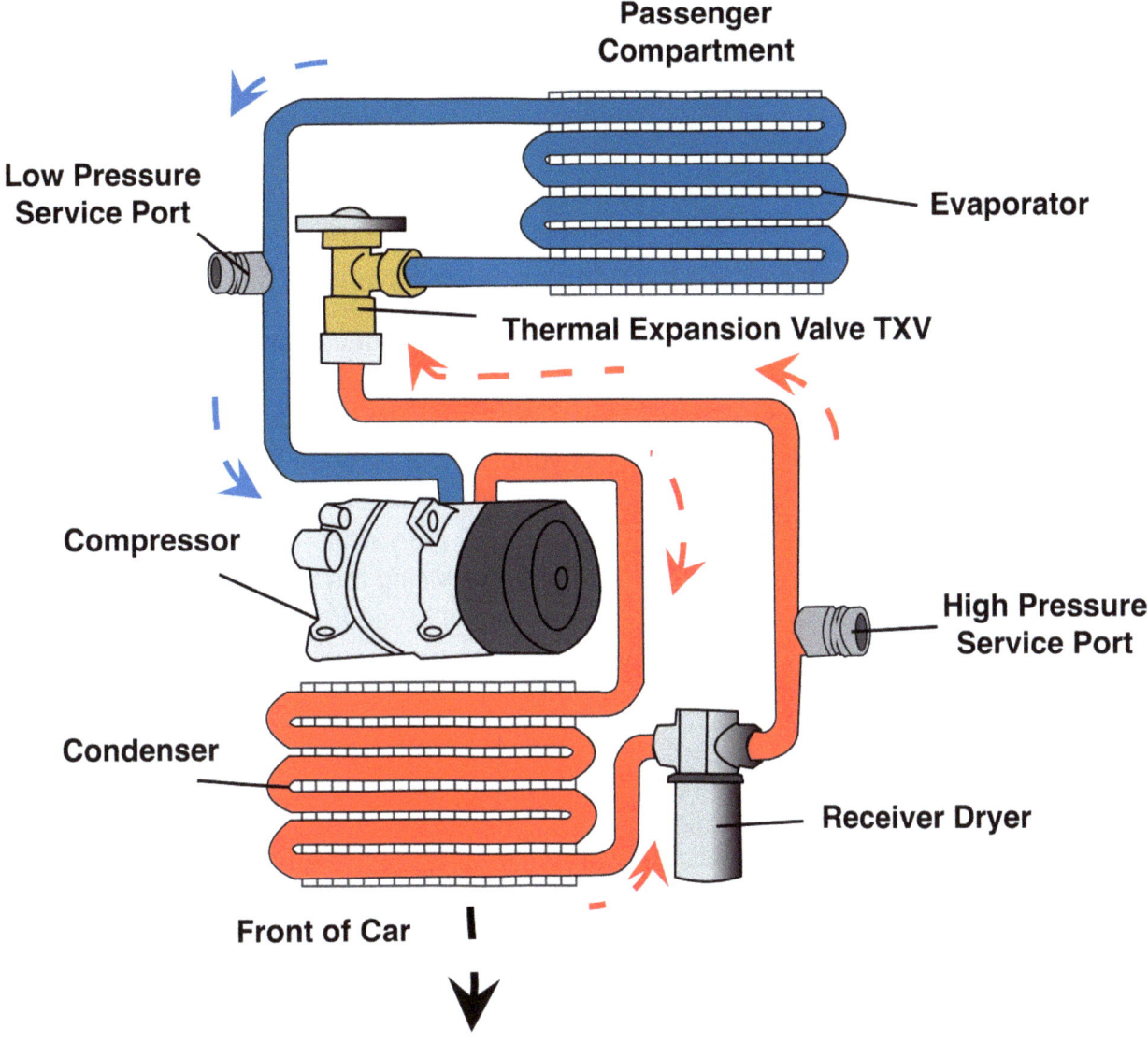

Figure 2.42 TXV air-conditioning circuit

An alternative to a TXV expansion valve system is the fixed orifice type. In this design, when the refrigerant leaves the condenser, it passes through an accurately sized restriction in the pipe work and sprays directly into the evaporator where it cools. When the refrigerant leaves the evaporator, any excess is stored in a suction accumulator before it is returned to the compressor.

Hybrid and electric vehicle air-conditioning and climate control

The air conditioning and climate control of a hybrid or electric drive vehicle works in the same way as a conventional system, with one main exception. Because the compressor in a conventional system is driven by the internal combustion engine, it cannot be operated when used with electric drive or during stop/start on hybrid motors. Instead the compressor itself is driven by an electric motor, powered from the high voltage batteries. Because the compressor uses the high voltage system, a special lubricating PAG (polyalkylene glycol) oil is needed that will not conduct electricity. Care must be taken when servicing the air conditioning systems of an electric or hybrid drive vehicle that only recommended oil is used.

The refrigerants used in air-conditioning systems are environmental pollutants. R12 creates ozone depletion and R134a contributes to the greenhouse effect. International agreements and protocols have led to legislation which restricts how refrigerants are used, and controls their release to atmosphere.

- ✓ List two types of internal combustion engine that can help improve fuel economy and lower emissions when used in a hybrid drive system.
- ✓ In relation to transmission systems, what do the letters 'CVT' stand for?
- ✓ What is special about the PAG oil used in the air conditioning systems of hybrid and electric vehicles?

Recommendations

At all stages of a diagnostic routine, maintenance or repair, you should record information and make suitable recommendations. The table below gives examples of how to do this.

Table 2.10 Recording information and making suitable recommendations

Stage	Information	Recommendations
Before you start	Record customer/vehicle details on the job card. Make a note of the customer's repair request and any issues/symptoms. Locate any service or repair history.	Advise the customer how long you will require the car. Describe any legal, environmental or warranty requirements.
During diagnosis and repair	Carry out diagnostic checks and record the results on the job card or as a printout from specialist equipment. List the parts required to conduct a repair. Note down any other non-critical faults found during your diagnosis.	Inform your supervisor of the required repair procedures so that they can contact the customer and gain authorisation for the work to be conducted.
When the task is complete	Write a brief description of the work undertaken. Record your time spent and the parts used during the diagnosis and repair on the job card. (This information should be as comprehensive as possible, because it will be used to produce the customers invoice). Complete any service history as required.	Inform the customer if the vehicle will need to be returned for any further work. Advise the customer of any other issues noticed during the repair.

Chapter 3 Alternative Fuel Vehicles

This chapter provides an overview of the function and operation of alternative fuel vehicles. It also describes the environmental issues created by traditional Petrol and Diesel. It will help you develop a systematic approach to the maintenance and repair of alternative fuel vehicles.

Contents

Safe working	129
Information sources	129
Electrical and electronic components	132
Environmental issues and low carbon technologies	132
The need for alternative propulsion	133
Exhaust emissions and pollution	136
Steam engines	143
Four-stroke petrol engines	144
Two-stroke petrol engines	147
Four-stroke Diesel engines	151
Two-stroke Diesel engines	153
Rotary engines	155
Liquefied petroleum gas (LPG)	158
Compressed natural gas (CNG)	162
Biogas	163
Biodiesel	164
Bioalcohol/Ethanol	168
Amonia Green	169
Solar	172
Mains supply (plug-in)	173
Hydrogen	174
Fuel cell	178
Alternative propulsion comparisons	180

Safe working when carrying out maintenance and repairs on alternative fuel vehicles

There are many hazards associated with the maintenance and repair of alternative fuel systems. You should always assess the risks involved with any inspection or repair routine before you begin and put safety measures in place. You need to give special consideration to the possibility of:

- Coming into contact with chemicals such as ammonia or acids. Many chemicals that are used to provide an alternative source of propulsion are highly toxic, and the corrosive or caustic nature can cause severe chemical burns.

- Pressurised containers and systems holding alternative fuels in liquid or gaseous form, which could cause fire, injury or death if accidentally damaged. (Although a pressure vessels comprising of a gas propulsion system fitted to a motor vehicle or trailer are exempt from the pressure systems regulations 2000).

Information sources

The complex nature of alternative vehicle propulsion systems requires you to have a good source of technical information and data. In order to conduct maintenance and repair procedures, you need to gather as much information as possible before you start. Sources of information include:

Table 3.1 Information sources

Information source	Example
Verbal information from the driver	A description of the symptoms produced which create an engine running problem, when the duel fuel vehicle is switched between petrol operation and LPG.
Vehicle identification numbers	Date of manufacture showing whether the LPG powered vehicle is E-OBD compliant (LPG vehicles became E-OBD compliant from 1st January 2005).
Service and repair history	Information on when the last service was conducted on a hydrogen powered internal combustion engine. This will help determine the condition of the engine oil due to water build-up caused by the combustion process (see HICE).
Warranty information	Has the conversion to alternative fuel propulsion affected the vehicle manufacturers warranty.

Table 3.1 Information sources

Vehicle handbook	Information on how to switch the fuel sources of a dual fuel vehicle between petrol and LPG.
Technical data manuals	A listing showing the recommended compression ratios used in a hydrogen powered internal combustion engine.
Workshop manuals	To find the recommended procedures to use when removing and refitting the hydrogen storage tank of a fuel cell powered vehicle.
Safety recall sheets	To confirm which components need to be replaced for the safe operation of a biodiesel internal combustion engine.
Manufacturer specific	Vehicle specific diagnostic trouble codes relating to the operation of a vehicles LPG system.
Information bulletins	Information on a common fault affecting vehicles running on bio alcohol.
Technical help lines	Advice on the correct routine for checking fuel pressure flow and volume.
Advice from master technicians/colleagues	An explanation of how to use the companies exhaust gas analyser when checking the emissions of a vehicle running on propane.
Internet	An Internet forum page where a number of people have had the same problems when converting their engines to run on biodiesel.
Parts suppliers/catalogues	A cross reference of spark plug part numbers to make sure that you use plugs with a lower heat range in a hydrogen powered internal combustion engine to prevent pre-ignition.
Jobcards	A general description on the work to be conducted on a customer's engine system.
Diagnostic trouble codes	A fault code showing that a lambda sensor (oxygen sensor) circuit needs to be tested to ensure correct engine management operation of a dual fuel vehicle.
Oscilloscope waveforms	A faulty current ramp waveform being produced from the fuel pump of a biodiesel powered vehicle.

Remember that no matter which information or data source you use, it is important to evaluate how useful and reliable it will be to your diagnostic, maintenance and repair routine.

Electronic and electrical safety procedures

When working with alternative fuel propulsion electrical and electronic systems, the main hazard is the possible risk of electric shock. Although most systems operate with low voltages of around 12 volts, an accidental electrical discharge caused by incorrect circuit connection can be enough to cause severe burns. Where possible, isolate electrical systems before conducting the repair or replacement of components.

When hydrogen fuel cells are in operation, they will create high voltages which are able to cause severe injury or even death. If you are working on hydrogen fuel cell vehicles, take care not to disturb the high voltage system. The high voltage system can normally be identified by its reinforced insulation and shielding which is often brightly coloured. (For more information on high voltage hazards and safety precautions, see Chapter 2).

Always use the correct tools and equipment. Damage to components, tools or personal injury could occur if the wrong tool is used or misused. Check tools and equipment before each use.

If you are using measuring equipment, always check that it is accurate and calibrated before you take any readings.

If you need to replace any fuel, electrical or electronic components, always check that the quality meets the original equipment manufacturer (OEM) specifications. (If the vehicle is under warranty, inferior parts or deliberate modification might make the warranty invalid. Also, if parts of an inferior quality are fitted, they might affect vehicle performance and safety.) You should only carry out the replacement of fuel or electrical components if the parts comply with the legal requirements for road use.

Table 3.2 The operation of electrical and electronic systems and components related to alternative fuel systems

Electrical/Electronic system component	Purpose
ECU	The electronic control unit is designed to monitor the operation of vehicle systems. It processes the information received and operates actuators that control system functions. An ECU may also be known as an ECM (electronic control module).
Sensors	The sensors are mounted on various system components and they monitor the operation against set parameters. As the vehicle is driven, dynamic operation creates signals in the form of resistance changes or voltage which are sent to the ECU for processing.
Actuators	When actioned by the ECU, motors, solenoids, valves, etc. help to control the function of vehicle for correct operation.
Digital principles	Many vehicle sensors create analogue signals (a rising or falling voltage). The ECU is a computer and needs to have these signals converted into a digital format (on and off) before they can be processed. This can be done using a component called a pulse shaper or Schmitt trigger.
Duty cycle and PWM	Lots of electrical equipment and electronic actuators can be controlled by duty cycle or pulse width modulation (PWM). These work by switching components on and off very quickly so that they only receive part of the current/voltage available. Depending on the reaction time of the component being switched and how long power is supplied, variable control is achieved. This is more efficient than using resistors to control the current/voltage in a circuit. Resistors waste electrical energy as heat, whereas duty cycle and PWM operate with almost no loss of power.

Environmental issues and low carbon technologies

As demand for natural resources rise, the impact on our environment also increases. One of the largest environmental pollutants is carbon dioxide, and **anthropogenic** (manmade) carbon dioxide is thought to be contributing to climate change.

Nearly everything we do in life creates carbon dioxide, and the amount of carbon dioxide produced through individual consumption and use is known as your **carbon footprint**. As the world's population increases, our requirements will also continue to rise and unless care is taken, we may cause irreparable damage to our planet.

Hybrid Electric & Alternative Automotive Propulsion

A large amount of carbon is generated through transport, and many manufacturers are exploring different technologies which can reduce overall CO_2 output (known as low carbon technologies), but there are also things that we can do as individuals when travelling that can help including:

- Ensuring that our vehicles are maintained and operate within recommended specifications
- Not making unnecessary journeys
- For short journeys, walk or cycle
- Using public transport where appropriate
- Sharing journeys with others (car sharing for example)
- Modify driving styles to ensure the most efficient operation of vehicles
- Use vehicles with alternative methods of propulsion

Anthropogenic - environmental pollution and pollutants originating from human activity.

Carbon footprint - the amount of carbon dioxide or other carbon compounds emitted into the atmosphere by an individual, company or country.

The need for alternative propulsion

In recent years a number of issues have arisen from the **fossil fuel** used for vehicle propulsion systems. These include:

- Limited quantities and reserves of crude oil
- Peak oil production
- Hazardous exhaust emissions
- Environmental pollution

Crude oil, which is the basis for all fossil fuels used in the propulsion systems of modern vehicles, is created by the biological breakdown of organic materials, from plant and animal life, underground. This process requires extreme heat and pressure, and takes place over millions of years. Crude oil then forms a rich source of hydrocarbons (HC) which can be extracted from the ground, and refined into fuels.

Low Carbon Technologies

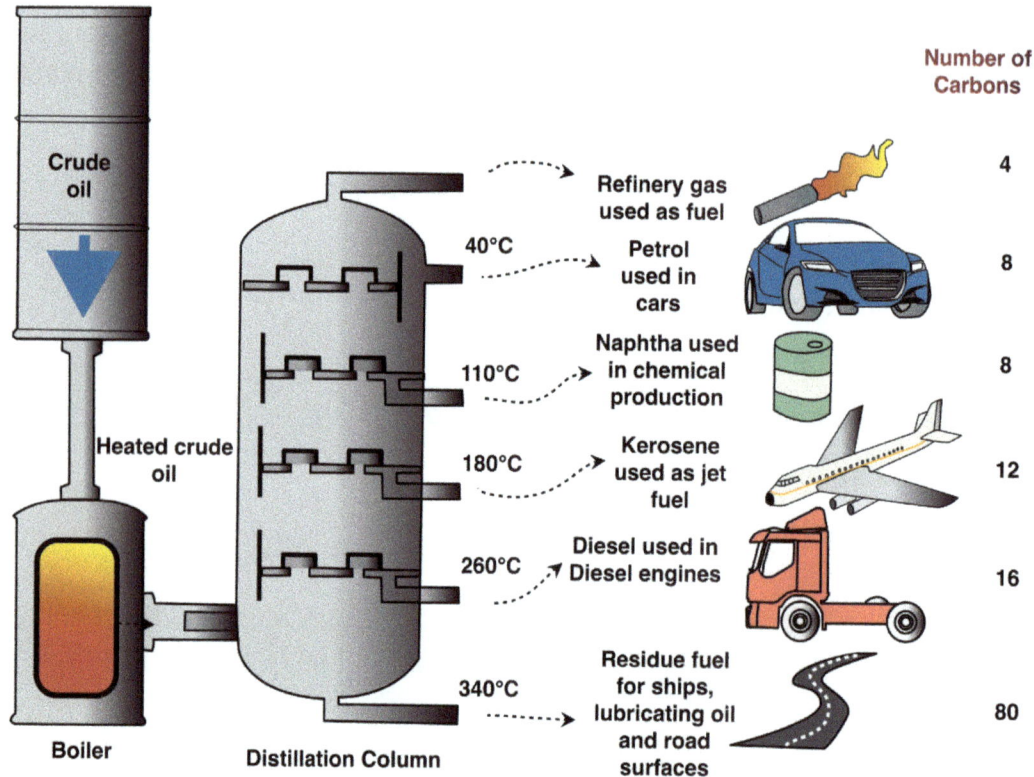

Figure 3.1 Refining crude oil

Petrol and Diesel fuels are made from crude oil. These fuels are distilled in an oil refinery, where the oil is heated and different chemicals boil off at various temperatures.

Whether a fuel is petrol or Diesel depends on how much hydrogen and carbon there are in the fuel. This is based on the number of carbon molecules in the hydrocarbon chain. The amounts can differ slightly depending on how the fuel is refined, but as a general rule:

- When the chain has between five and nine carbons, the hydrocarbon is petrol (gasoline).

- When the chain has about 12 carbons, the hydrocarbon is Diesel.

Figure 3.2 Petrol and Diesel hydrocarbon chains

Each fuel has a **calorific value**, which is the energy stored within it. (This is measured in the same way as the calories in food). When the fuel is burnt, this energy is given up in the form of **thermal energy**, which can be used to provide motion power to the vehicle. The internal combustion engine is not very good at converting heat stored into other forms of energy. As a result, only about 20% of the energy produced by burning petrol is used. This means that 80% of the stored thermal energy is wasted.

Diesel is slightly more efficient, with a theoretical energy use of around 73%. This is why a car can travel so much further on a litre of Diesel than it could on the same quantity of petrol.

Fossil fuel – naturally occurring oil and fuels created from plant and animal life that has decayed over millions of years.

Thermal energy – energy stored as heat.

Calorific value – the amount of energy stored within a fuel.

The amount of energy given up when petrol or Diesel is burnt inside an engine is greater than that of dynamite.

Limited quantities and reserves of crude oil

Unfortunately the amount of crude oil that is available is limited. Millions of barrels of oil are extracted from the ground every day, meaning that this vast energy supply is rapidly being used up. Once all the known crude oil has been extracted from the ground, there will be no more to replace it for millions of years to come.

Peak oil production

Another factor affecting the use of crude oil as a reliable fuel source, is a situation known as 'peak oil production'. As requirements for power in the modern world have increased, more and more oil needs to be extracted and processed to keep up with demand. Peak oil is a speed limit barrier with which this can be done, and fuel shortages will arise when demand outpaces supply.

Exhaust emissions

When a fossil fuel is burnt inside an engine, it is known as combustion.

Three things are needed for combustion:

- a source of fuel
- a source of heat
- a source of oxygen

With fossil fuels the source of fuel is either the petrol or Diesel, but it is only the hydrogen in the hydrocarbon fuel that is burnt. This means that the carbon that is found in petrol and Diesel is a waste product.

Figure 3.3 Exhaust emissions

The source of oxygen that is used for combustion comes from air.

The constituent components of air are:

- nitrogen at 78%
- oxygen at 21%
- 1% other gases

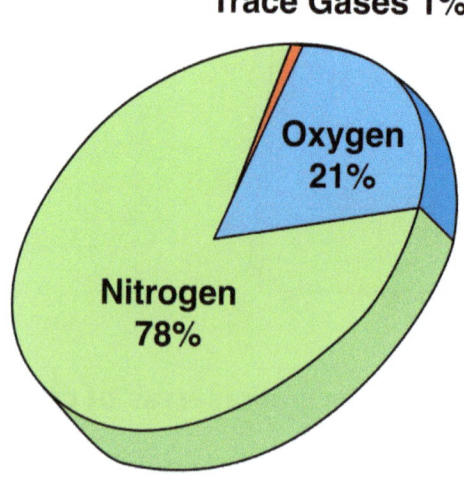

Figure 3.4 The main constituents of air

As it is only the oxygen that is used in the combustion process, the nitrogen is a waste product. Nitrogen is an inert gas, which means that it doesn't burn or support combustion.

When oxygen and hydrogen are brought together and are supplied the source of heat via the spark at the plug or the superheated air created by compression, they burn and release their energy as heat.

In order to minimise the amount of waste products given off by the combustion process, just the right ratio of oxygen to hydrogen is required so that after combustion, very little is left over. This is known as a **stoichiometric value** or a balanced chemical reaction. Under normal operating conditions the stoichiometric value should be 14.7:1 by mass (14.7 parts air to 1 part fuel, by weight).

Some exhaust pollutants are hazardous to health.

When working on a running engine you should avoid breathing in exhaust fumes by using special exhaust fume extraction equipment. If this is not available you should make sure that the area is well-ventilated.

Exhaust gas constituents

Exhaust gases consist of chemical components made up after the combustion process has taken place.

The main chemical elements taken into the engine are: oxygen and nitrogen in air, and hydrogen and carbon in fuel.

Theoretically, ideal combustion (sometimes called the **Lambda window**) would only produce carbon dioxide, nitrogen and water (H_2O). Unfortunately, the combustion process is never perfect and as a result exhaust pollutants are produced.

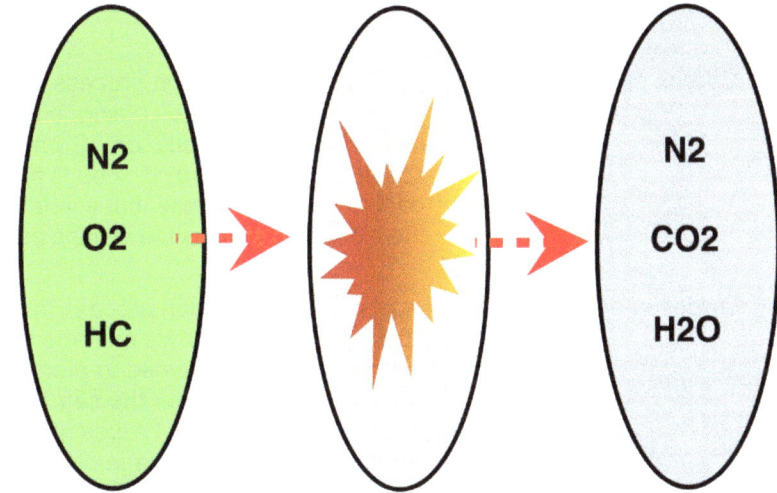

Figure 3.5 Ideal combustion

Following the combustion process of fossil fuels the elements are rearranged to form other substances as described in table 3.3 which are considered pollutants.

Stoichiometric values – the ideal air to fuel ratio – approximately 14.7 to 1 by mass.

Lambda window – the engine operating range which achieves an air to fuel ratio of 14.7:1 by mass.

Lambda – a Greek letter used to represent the ideal air to fuel ratio or stoichiometric value.

Table 3.3 Exhaust emissions and their descriptions

Exhaust emission	Description
Carbon dioxide (CO_2)	With effective combustion, the chemical elements carbon and oxygen found in the fuel and air, combine to form carbon dioxide. CO_2 emissions should normally be higher than 14% by volume. Carbon dioxide is a greenhouse gas and is considered an environmental pollutant.
Hydrocarbons (HC)	If fuel passes through the combustion process with no chemical change, hydrocarbons are given off. These should be kept as low as possible, usually under 200 parts per million at idle. Hydrocarbons are considered harmful to health and may cause lung damage or cancer.
Carbon monoxide (CO)	If during the combustion process, the burning fuel goes out (maybe due to lack of oxygen or rapid cooling, also known as **quenching**), carbon monoxide (CO) is produced. Carbon monoxide is a product of incomplete combustion. Carbon monoxide is harmful to health. It is colourless, odourless and tasteless, but if inhaled it is poisonous. It replaces oxygen in the blood and starves the organs of required oxygen.
Oxides of nitrogen (NOx)	Most of the combustion process takes place at temperatures between 2000 and 2500°C. In this extreme heat, the oxygen and nitrogen in the incoming air are combined to produce a pollutant called oxides of nitrogen. Unfortunately, the better the combustion process is, the more oxides of nitrogen are produced. Oxides of nitrogen can cause health issues for the lungs and may also damage plant life and reduce visibility.

Hybrid Electric & Alternative Automotive Propulsion

Quenching - the rapid cooling of combustion, leading to the ignited fuel being extinguished.

Exhaust emission standards

Exhaust emission standards are continually being updated and revised, with Euro 4 being the most common standard currently. The stricter Euro 5 regulations are now compulsory for all new cars currently on sale in the UK.

The European emission standards are set out in Table 3.5. They show the limits set in grams per kilometre (g/km).

Table 3.5 European emission standards

Standard	Commencing	Carbon monoxide limits	Total Hydrocarbons	Non-Methane Hydrocarbons	Oxides of Nitrogen	Hydrocarbons plus Oxides of nitrogen	Particulate matter (soot)
Limits for Diesel engine cars							
Euro 1	July 1992	2.72 (3.16 COP)	N/A	N/A	N/A	0.97 (1.13 COP)	0.14 (0.18 COP)
Euro 2	January 1996	1.0	N/A	N/A	N/A	0.7	0.08
Euro 3	January 2000	0.64	N/A	N/A	0.50	0.56	0.05
Euro 4	January 2005	0.50	N/A	N/A	0.25	0.30	0.025
Euro 5	September 2009	0.500	N/A	N/A	0.180	0.230	0.005
Euro 6	September 2014	0.500	N/A	N/A	0.080	0.170	0.0025

Table 3.5 European emission standards

Limits for petrol engine cars							
Euro 1	July 1992	2.72 (3.16 COP)	N/A	N/A	N/A	0.97 (1.13)	N/A
Euro 2	January 1996	2.2	N/A	N/A	N/A	0.5	N/A
Euro 3	January 2000	2.3	0.20	N/A	0.15	N/A	N/A
Euro 4	January 2005	1.0	0.10	N/A	0.08	N/A	N/A
Euro 5	September 2009	1.000	0.100	0.068	0.060	N/A	0.005
Euro 6	September 2014	1.000	0.100	0.068	0.060	N/A	0.005
OP = Conformity of Production							

Using an exhaust gas analyser

How to:

1. Switch on the exhaust gas analyser and allow it to go through its pre-set warm-up and calibration procedure.
2. Depending on the fuel type being measured, some exhaust gas analysers have the ability to be set up to accommodate petrol, LPG and CNG. Following manufacturer's instructions, set analyser to correct fuel type.
3. Connect the exhaust extraction equipment, start the engine and run up to normal operating temperature.
4. Insert the exhaust probe at the tail pipe.
5. Allow the digital reading to settle on the analyser (this may take up to 15 seconds) and compare with the manufacturer's recommendations.

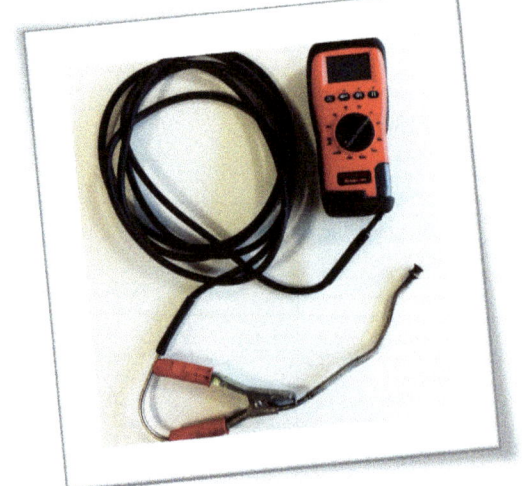

Figure 3.6 A handheld exhaust gas analyser

If you are running the engine in the workshop, you must use exhaust extraction equipment to protect you from harmful pollutants. Many exhaust extraction units include a specially designed hole for use with the exhaust gas analyser probe.

If your exhaust gas analyser does not have the capability of measuring LPG or CNG, the results form a standard petrol analysis check may be used with a calculation to correct the values of carbon monoxide. On the back of some analysers a sticker will show the values to be used for this calculation. They will often be listed as PEF (propane equivalency factor) or MEF (methane equivalency factor). Depending on the type of fuel used the carbon monoxide results should be multiplied by the equivalency factor to obtain the corrected readings.

Environmental pollution

A by-product of the combustion process when burning any fossil fuel, is the release of carbon dioxide. In fact the more efficient the combustion process is, the more CO_2 is produced. Carbon dioxide is a greenhouse gas, and thought to contribute to global warming.

Global warming

Most of the planets energy comes from the sun in the form of electromagnetic radiation, and one component of this is visible light. Because of its wavelength, light is able to travel easily through the atmosphere and reach earth where its energy is absorbed by the ground. At night when this energy is given up by the ground, it tries to radiate back to space as infrared radiation. Unfortunately, the wavelength of infrared has a slightly longer frequency than visible light, and greenhouse gasses trap this radiation leading to global warming. The gasses in our atmosphere are classified by their ability to trap infrared radiation and given a number to represent their Global Warming Potential (GWP). The Global Warming Potential of carbon dioxide is very low and it is given a GWP of 1. The reason why carbon dioxide is such an environmental issue, is the shear amount of it that is produced by mankind (known as anthropogenic CO_2). A large proportion of anthropogenic CO_2 comes from the burning of fossil fuels used in vehicles for transport.

Low Carbon Technologies

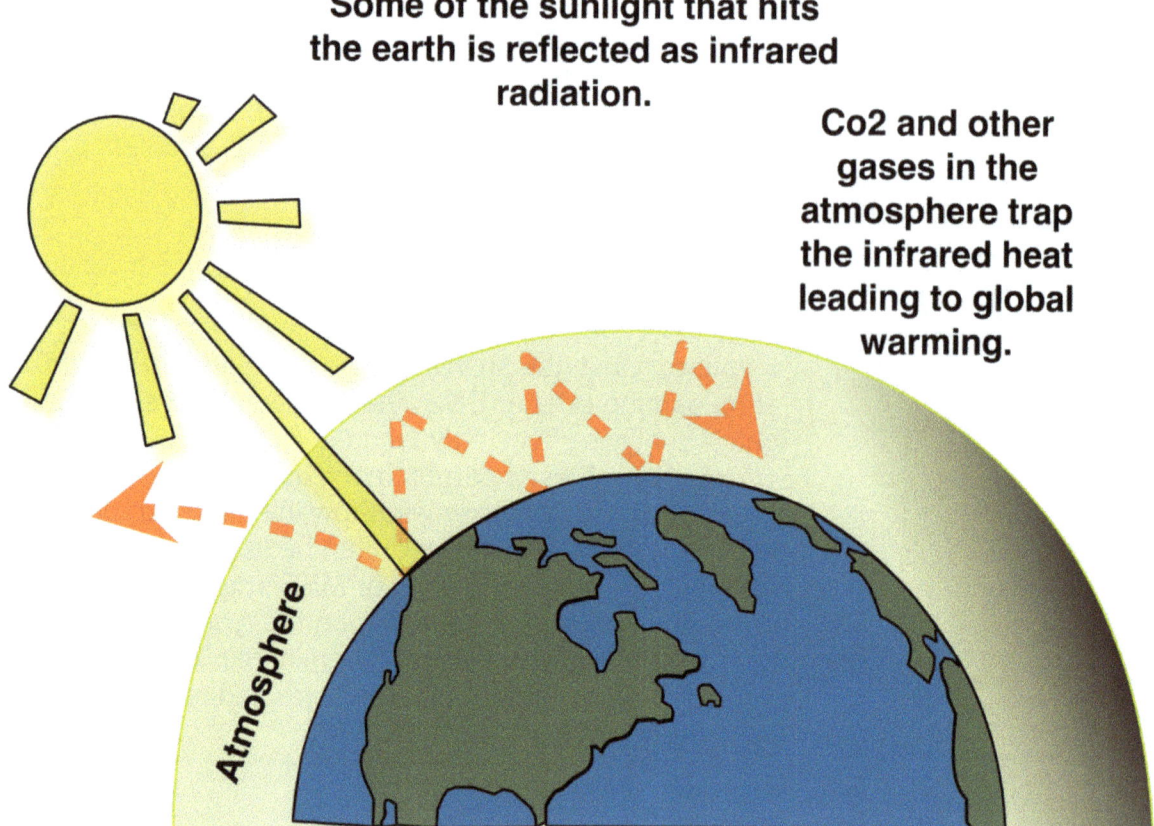

Figure 3.7 Infrared radiation causing global warming

To try and overcome the issues created by fossil fuel use, vehicle manufacturers are investigating alternative methods of propulsion.

- What is a fossil fuel?
- List four exhaust pollutants produced from burning fossil fuels.
- What do the letters 'GWP' stand for?

Engine types and operation

External combustion engines

Early vehicles used external combustion engines (steam engines) to provide propulsion. Although steam is no longer employed for everyday use, a number of vehicle designers have created steam cars as engineering exercises and produced high performance output.

Water can be heated in a boiler, powered by the combustion of a number flammable materials (although some of these are fossil fuels). If the water system is highly pressurised, it is able to produce superheated steam which can then be used to drive pistons or turbines and create propulsion. The disadvantage of using steam as a method of propulsion is that a tank needs to be constantly maintained with a supply of water, in order for the vehicle to move.

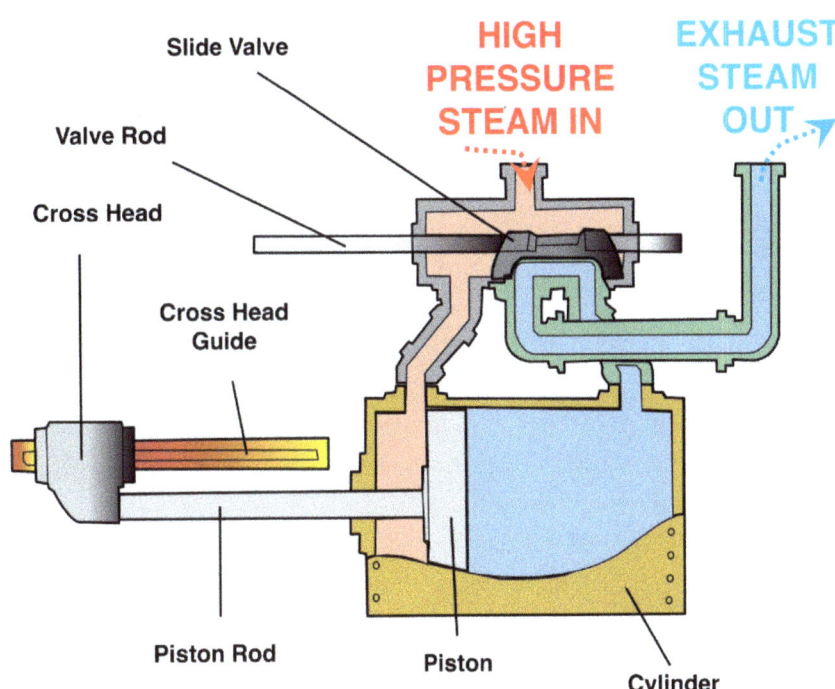

Figure 3.8 Steam engines

Internal combustion engines

The internal combustion engine has been around for over one hundred years and is based on a design by Nicolaus August Otto. He was the first person to develop an engine that made use of four separate cycles of operation, and in 1876 he produced a gas powered engine that he fitted to a motorcycle.

The Otto cycle is the most commonly used method for powering the internal combustion engines used in cars, lorries and motorcycles today.

Spark ignition and compression ignition

There are two main categories of Otto cycle: **spark ignition (SI)** and **compression ignition (CI)**.

Low Carbon Technologies

The operating cycle for four- and two-stroke engines

The four-stroke operating cycle of a spark ignition (SI) petrol engine

The four-stroke cycle of operations is used in **reciprocating** engines.

> **Spark ignition (SI)** – petrol engines, which need a spark to ignite the fuel.
>
> **Compression ignition (CI)** – Diesel engines, which use hot compressed air to ignite the fuel.
>
> **Reciprocating** – an engine that uses pistons moving backwards and forwards or up and down.
>
> **Cylinder bore** – the area inside the engine that houses the pistons.

Induction (Suck)

As the piston moves downwards in the **cylinder bore**, a low pressure (also called a 'depression') is created above it in the combustion chamber.

The camshaft turns and operates on a mechanism to open the inlet valve.

The atmospheric pressure pushes the air/fuel mixture through the open inlet valve and tries to fill the cylinder.

As the piston reaches the bottom of the induction stroke, the lobe of the camshaft moves away from the valve-operating mechanism, and a return spring is used to close it. This seals the combustion chamber.

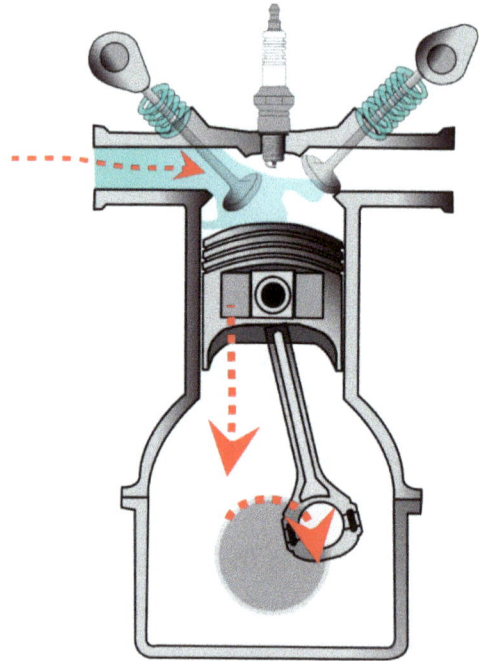

Figure 3.9 The induction stroke of the four-stroke cycle

Hybrid Electric & Alternative Automotive Propulsion

Vaporise – to turn liquid fuel into vapour or gas.

Compression (squeeze)

As the crankshaft continues to turn, the piston moves upwards, squeezing the air/fuel mixture into a smaller and smaller space, which raises its pressure. This rise in pressure will perform two functions:

1. As the pressure rises, the temperature of the air/fuel mixture increases. This rise in temperature helps to **vaporise** the fuel. This is important because it is the fumes (vapour) of the petrol that actually burn, not the liquid.

2. The rise in pressure means the air/fuel mixture is burnt in the confined space of the combustion chamber above the piston. The fuel releases its energy in a much more powerful manner than if it was not confined.

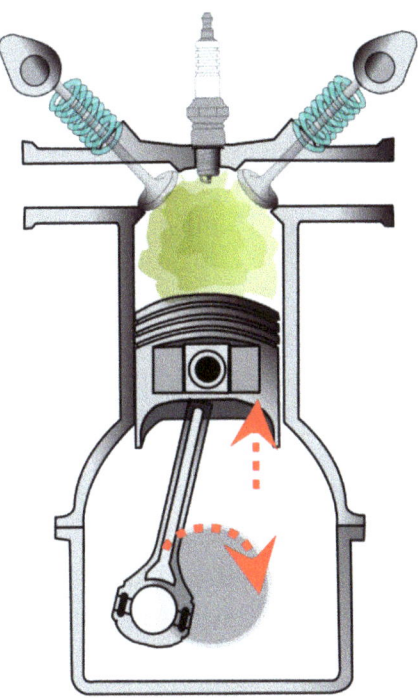

Figure 3.10 The compression stroke of the four-stroke cycle

When a fuel is rapidly burnt in a confined space, it releases its energy in a much more powerful manner. For example, if you set light to a small pile of gunpowder, it will burn away very quickly. But if you take the same gunpowder and seal it inside a cardboard tube, it will explode in the same way as a firework.

Just before the piston reaches the top of the compression stroke, a spark plug mounted in the combustion chamber releases a high-voltage spark. The heat from the spark is enough to start the air/fuel mixture burning. This creates a flame that spreads out evenly as it moves away from the spark plug. There is a rapid expansion of gases created by the heat.

Power (bang)

Because the gases are sealed in the confined space of the combustion chamber above the piston, there is a rapid pressure rise. This forces the piston downwards on a power stroke.

The energy from the power stroke is transferred through the piston and connecting rod to the crankshaft, which then turns.

The rotation of the crankshaft on the power stroke is the only active operation within the Otto cycle. The other three strokes are often called 'dead' strokes because they do not help with the turning of the crankshaft.

Figure 3.11 The power stroke of the four-stroke cycle

Kinetic energy stored in the flywheel is used to keep the crankshaft turning on the dead strokes.

Exhaust (blow)

As the piston reaches the bottom of the power stroke, the energy in the burning air/fuel mixture is used up. This leaves waste products behind, which are known as exhaust gases.

The exhaust valve is opened by the camshaft, and as the piston travels upwards, the exhaust gases are forced out through the **manifold** and exhaust system, entering the atmosphere as pollutants.

As the piston reaches the end of the exhaust stroke, the valve is closed by the return spring and the inlet valve begins to open. The whole cycle of operations starts again.

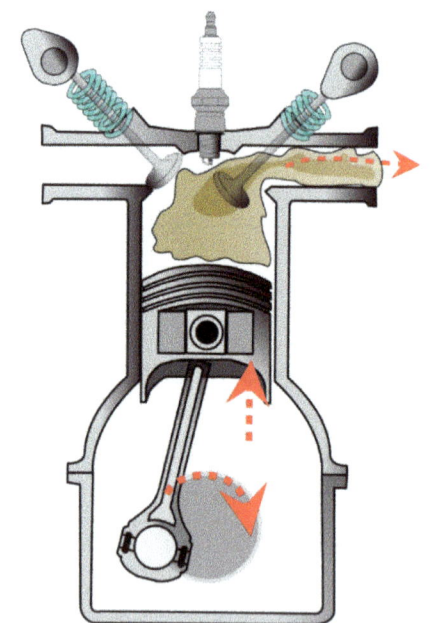

Figure 3.12 The exhaust stroke of the four-stroke cycle

Manifold – a set of pipes used to supply air and fuel to the engine or to direct exhaust gases away.

Crankshaft rotation in the four-stroke cycle of operations

To complete one full cycle of operations (induction, compression, power, exhaust), the crankshaft of a standard petrol engine must make two full revolutions and the piston will move up and down twice (two strokes each revolution equals four strokes in total). This will give 720° of crankshaft rotation.

The two-stroke operating cycle of a spark ignition (SI) petrol engine

Two-stroke engines are rarely used in cars because in the past they have been inefficient and highly polluting. However, advances in technology have led to improvements in the operating principles. As a result, some engine manufacturers are considering reintroducing this engine type.

A two-stroke engine completes a full cycle of operation in just two strokes (the piston moves up and down once). With a four-stroke engine, the entire process takes place above the piston, but a two-stroke engine also makes use of the area below the piston (the crankcase) to speed up the operation.

Instead of valves, most two-stroke engines use **ports** that are opened and closed (covered and uncovered) by the piston as it moves up and down.

Ports – holes machined in the cylinder walls.

When you are reading the description of the two-stroke operating cycle that follows, try to imagine what is going on both above and below the piston at the same time.

Induction

Induction takes place below the piston. As the piston moves upwards, a low pressure is created in the crankcase. The piston uncovers the inlet port and an air/fuel mixture is drawn into the crankcase.

Some two-stroke engines use a passive valve or flap called a 'reed valve' to help seal the crankcase once the air/fuel mixture has been inducted.

Pre-compression

As the piston moves downwards, it closes the inlet port and squeezes the air/fuel mixture in the crankcase, raising its pressure.

Transfer

As the piston continues downwards, it opens the **transfer port** in the cylinder wall. The pressurised mixture is forced upwards above the piston. (At this point, an exhaust port machined in the cylinder wall opposite the transfer port is also open).

Compression

The piston moves upwards (covering both the transfer and exhaust ports) and compresses the air/fuel mixture into the combustion chamber.

Power

Just before the piston reaches top dead centre (TDC), the spark plug ignites the air/fuel mixture and the high pressure created forces the piston downwards on its power stroke.

Exhaust

As the piston moves downwards on its power stroke, it uncovers the exhaust port, and the fresh incoming mixture from the transfer port helps to push the burnt gases out of the engine. This process is called **scavenging**.

Hybrid Electric & Alternative Automotive Propulsion

Notes: In a standard two-stroke engine, lubrication of the crankcase components is achieved by mixing oil with the inducted air/fuel as a fine mist. This means that the lubrication oil is burnt during the combustion process, adding to exhaust pollution. Because the oil is consumed by the engine, and requires constant topping up, this is known as a total loss lubrication system.

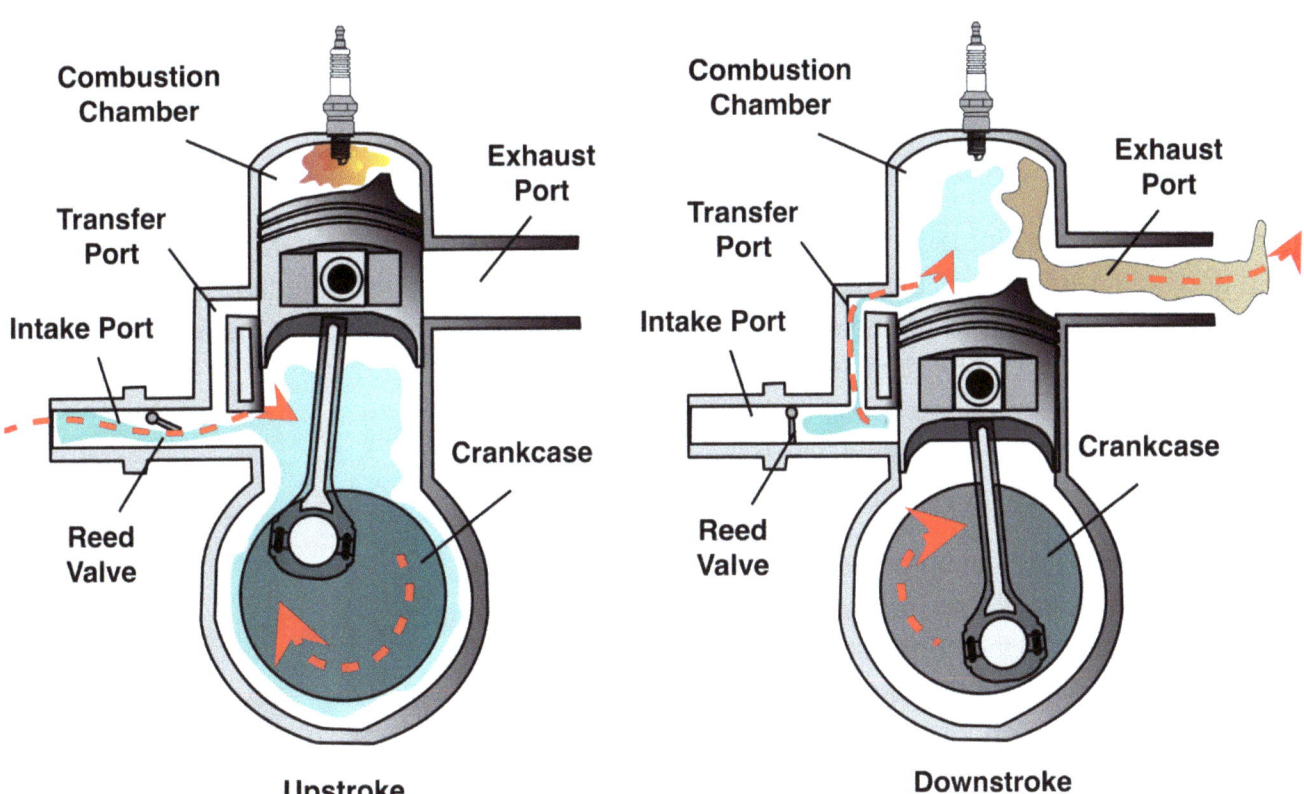

Figure 3.13 The two-stroke petrol engine cycle of operations

Low Carbon Technologies

Transfer port – a passageway from the crankcase to the combustion chamber that is machined in the cylinder wall.

Scavenging – assisting the removal of exhaust gases using pressure from the incoming air/fuel mixture.

Crankshaft rotation in the two-stroke cycle of operations

To complete one full cycle of operations (induction, pre-compression, transfer, compression, power and exhaust), the crankshaft of a two-stroke petrol engine only makes one full revolution and the piston will move up and down once (two strokes in total). This will give 360° of crankshaft rotation.

Low pressure direct injection two-strokes

Instead of having the air/fuel mixture drawn into the crankcase as on a standard two-stroke, some engine manufacturers are using a low pressure air and fuel direct injection systems in their engine designs. In this type of engine lubrication oil is pumped directly to the moving crankcase components. As the amount of oil used is precisely controlled, oil consumption during the combustion process is reduced which lowers exhaust emissions and pollution.

Air supply for the combustion process still comes via the crankcase and transfer, but fuel is delivered directly to the combustion chamber from an injector. A crankshaft-driven compressor is able to supply pressurised air to the electronic fuel injector. As the injector operates, fuel is **atomised** by the compressed air and only injected after the exhaust port has closed. This prevents the fuel escaping from the exhaust port during scavenging.

The spark plug ignites the air/fuel mixture to produce the power stroke and the whole process still only takes one revolution of the crankshaft (two strokes).

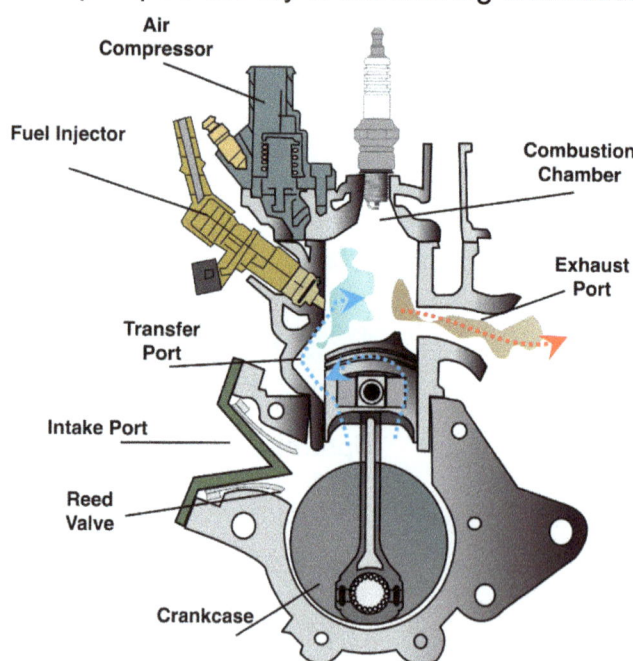

Figure 3.14 Low pressure two stroke petrol engine

Atomise – to break up liquid fuel into small droplets or a spray.

Comparison of two-stroke and four-stroke components

Although two-stroke and four-stroke reciprocating engines both use pistons and crankshafts to operate, many of the other components required for a four-stroke to function are not needed in a two-stroke. In fact around three-quarters of the moving parts (valves, valve trains, etc) can be removed. This means that if a two-stroke cycle is used, lighter, quicker engines can be constructed.

The operating cycle of a compression ignition (CI) Diesel engine

The four main elements of the Otto cycle are still used in a compression ignition (CI) engine, but the induction and the ignition processes are different from those found in a spark ignition (SI) engine.

In 1892 Rudolf Christian Karl Diesel took out a patent for his design of compression ignition, and his name has become associated with these engines.

Stronger and heavier mechanical components are needed in a CI engine because of the higher pressures and stresses involved, but compression ignition is a more efficient way of generating energy from fuel than spark ignition. This means that a CI engine will run longer on Diesel than a SI engine will run on the same amount of petrol.

Induction

The induction stroke of a CI engine is similar to that of a SI engine. (See the explanation of the SI induction (suck) process). But there is a key difference:

- In the induction stroke of a CI engine, only air is drawn into the cylinder; no fuel is mixed with the air at this stage.

Compression

The compression stroke of a CI engine is similar to that of a SI engine. (See the explanation of the SI compression (squeeze) process). But there are a number of key differences:

- A much smaller combustion chamber is used in a CI engine than in a SI engine.

- Because of the higher pressure inside a CI engine, the air becomes superheated and this is used as the source of ignition instead of a spark plug.

- Just before the piston reaches the top of the stroke, a fuel injector mounted in the combustion chamber (or a pre-combustion chamber, sometimes called a 'swirl chamber') opens and sprays Diesel fuel directly into the superheated air. As the small droplets of Diesel fuel come into contact with this superheated air, they spontaneously combust (without the need for a spark plug) and begin to burn.

- To try to produce a smooth burn from the air/fuel mixture, the combustion chamber and/or piston are shaped to create turbulence (also known as swirl), which attempts to mix the Diesel and air evenly.

Power

The power stroke of a CI engine is similar to that of a SI engine. (See the explanation of the SI power (bang) process). But there is a key difference:

- The droplets of Diesel injected into the superheated air begin to burn, and the heat created makes the gases expand. Unlike a petrol engine, where the burn spreads out evenly away from the spark plug, Diesel droplets begin to burn in a number of different positions and directions. As these burning droplets (called flame fronts) collide, a noise sometimes called 'Diesel knock' is created.

Exhaust

The exhaust stroke of a CI engine is the same as that of a SI engine. (See the explanation of the SI exhaust (blow) process).

Crankshaft rotation in the standard CI engine cycle of operations

To complete one full cycle of operations (induction, compression, power, exhaust), the crankshaft of a standard CI Diesel engine must make two full revolutions and the piston will move up and down twice (two strokes each revolution equals four strokes in total). This will give 720° of crankshaft rotation.

Hybrid Electric & Alternative Automotive Propulsion

The operating cycle of a two-stroke compression ignition (CI) Diesel engine

The two-stroke Diesel engine is not currently used in production cars, but is in widespread use for heavy vehicles and marine engines. This type of engine is highly efficient at extracting thermal energy from fuel, which makes it an interesting concept for engine and vehicle designers.

Induction

The induction stroke of a two-stroke CI engine is similar to that of a four-stroke CI engine. (See the explanation of the CI induction process). But there are some key differences:

- As the piston moves downwards, it uncovers ports in the cylinder wall. No inlet valve is required.

- Some form of **supercharger** or **compressor** is needed to push the air into the cylinder through the open ports in the wall.

- As the piston reaches the bottom of the induction stroke, the piston starts to move back up the cylinder, closing off the intake ports and sealing the combustion chamber.

Supercharger/Compressor – a mechanical device that can be used to 'blow' air into the engine.

Compression

The compression stroke of a two-stroke CI engine is the same as that of a four-stroke CI engine. (See the explanation of the CI compression process).

Power

The power stroke of a two-stroke CI engine is the same as that of a four-stroke CI engine. (See the explanation of the CI power process). As with a four-stroke Diesel engine, the noise known as 'Diesel knock' is often produced.

Exhaust

The exhaust stroke of a two-stroke CI engine is similar to that of a four-stroke CI engine. (See the explanation of the CI exhaust process). But there are some key differences:

- The inlet ports in the cylinder wall will have been uncovered by the piston, following the downwards power stroke. When the exhaust valve is opened by the camshaft, the high air pressure created by the supercharger forces the waste gas out through the open exhaust valve and on through the manifold and exhaust system.

- As the piston reaches the end of the exhaust stroke, the supercharger has forced in a fresh charge of air to start the cycle again.

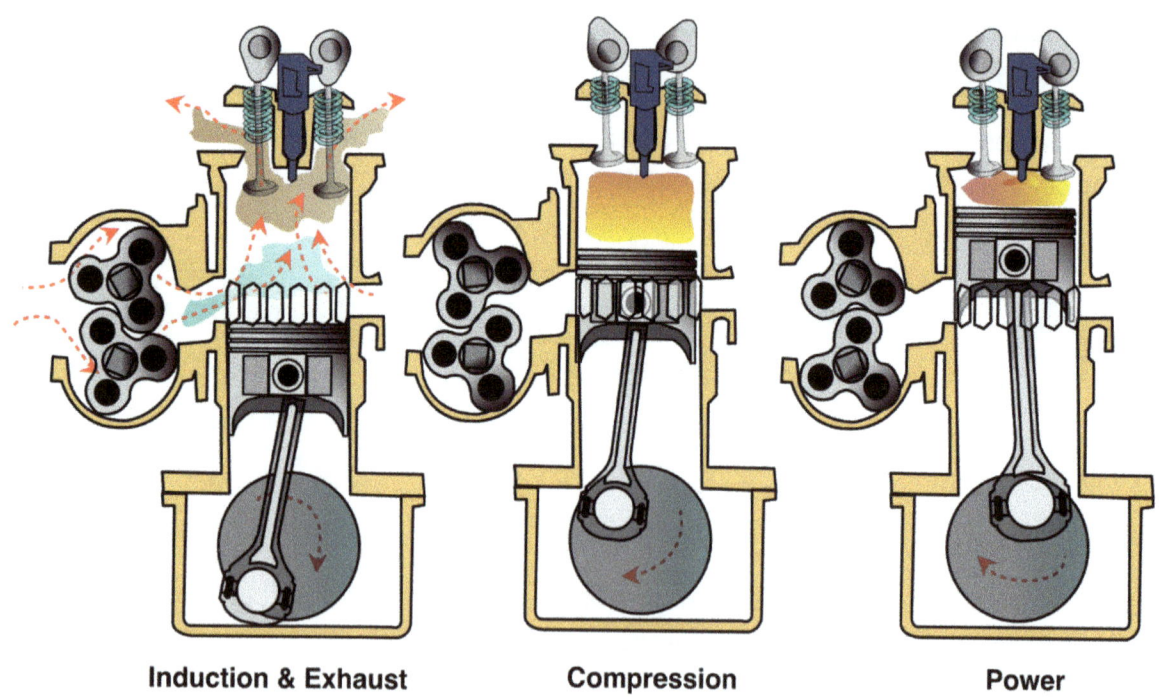

Figure 3.15 A two-stroke compression ignition engine

Crankshaft rotation in the two-stroke Diesel engine cycle of operations

To complete one full cycle of operations (induction, compression, power, exhaust), the crankshaft of a two-stroke Diesel engine only makes one full revolution and the piston will move up and down once (two strokes in total). This will give 360° of crankshaft rotation.

Comparison between engine components

Spark ignition SI (petrol) and compression ignition CI (Diesel) engine components are similar in operation, but built differently. Because Diesel engines work with much higher pressures, the mechanical construction of their components needs to be more robust. This means that Diesel engine components are normally much heavier. When heavy components are used inside an engine, they are unable to rev as high.

The operating cycle of a rotary engine

Rotary engines are a form of SI petrol engine that do not use a conventional piston. The type of rotary engine used in many modern cars is based on a design made by a German mechanical engineer, Felix Wankel, in the 1950s.

A rotary engine is different from a standard reciprocating (up and down) engine because it does not have a normal piston. Instead of a piston, it has a three-sided **rotor** sitting inside a squashed oval-shaped cylinder called an 'epitrochoid'.

Because the rotor has three sides, it is effectively three pistons in one. This means that a different **phase** of the Otto cycle (induction, compression, power or exhaust) will be happening on each side of the rotor at the same time.

Unlike a standard four-stroke engine, the rotary engine doesn't have valves. Instead, it has ports in the cylinder wall that are opened and closed as the rotor turns.

Rotor – a triangular component used instead of a piston in the rotary engine.

Phase – a term used instead of 'stroke' when describing the operation of a rotary engine.

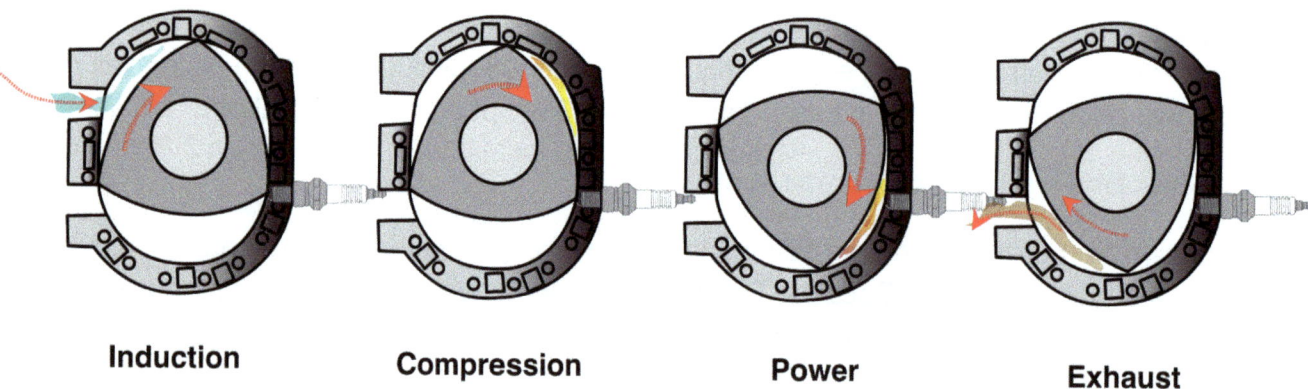

Figure 3.16 The four phases of a Wankel rotary engine

Induction

As the rotor turns, the tip uncovers the inlet port. An expanding chamber is produced which creates the low pressure (or depression). Atmospheric pressure or a turbocharger forces an air/fuel mixture through the open inlet port and tries to fill the cylinder. As the rotor reaches the end of the induction phase, the tip covers the inlet port, sealing the combustion chamber.

Compression

As the rotor continues to turn, the sealed chamber starts to reduce in size, compressing the air/fuel mixture into a smaller and smaller space, which raises its pressure. This rise in pressure performs two functions:

1. As the pressure rises, the temperature also increases. This rise in temperature helps to vaporise the fuel. This is important because it is the fumes or vapour of the petrol that actually burn, not to the liquid.

2. When a fuel is rapidly burnt in a confined space, it releases its energy in a much more powerful manner.

Power

At the point of highest compression, a spark plug ignites the air/fuel mixture. Because these gases are sealed in a rotor chamber, a rapid pressure rise occurs, forcing the rotor round on a power phase.

The energy from the power phase is transferred through the rotor to the crankshaft, which is turned.

The rotation of the crankshaft on the power phase is the only active operation within the Otto cycle, but as the rotor has three sides, it is quickly followed by another power phase.

This means that a smaller, lighter flywheel can be used than in a standard reciprocating engine. The crank produces a lower amount of **torque** than a standard piston engine, but can give a high power output.

Exhaust

At the end of the power phase, the rotor tip uncovers the exhaust port. The chamber continues to decrease in size, forcing the burnt exhaust gases out through the manifold and exhaust system. At this point the process repeats itself.

Torque – turning effort. In this case, it is the turning effort produced at the crankshaft as it is rotated. The torque will normally increase the amount of weight a vehicle can pull.

Crankshaft rotation in the rotary engine cycle

Because the rotary engine has a rotor with three sides, there is a power phase every 120° of revolution. Many rotary engines combine two rotor chambers and will provide a power phase every 60°.

- In the Otto cycle, what two functions does the compression stroke perform?
- What is the name for the oval shaped cylinder in a rotary engine?
- In a low pressure direct injection two-stroke engine, how is oil delivered to crankcase components?

Alternative fuel vehicles

Petrol and Diesel are gradually becoming more expensive to produce. One of the main reasons for this is that the demand for these fuels has increased. To overcome the issues created by 'peak oil' and environmental pollution many manufacturers are researching and designing cars that will run on alternative fuel.

An alternative fuel vehicle is a vehicle that runs on a fuel other than traditional petroleum-based fuels (i.e. petrol or Diesel). Between 2008 and 2009 there were around 33 million alternative fuel and advanced propulsion technology vehicles on the world's roads. This represented around five per cent of the world's vehicles. This number is increasing every year.

Liquefied petroleum gas (LPG)

LPG, sometimes known as 'autogas' is produced during the normal refining process of crude oil. Liquefied petroleum gas (LPG) is a low pressure liquefied gas mixture made up of mainly propane and butane, but essentially it is petrol gas. This fuel can be burned in a conventional internal combustion engine and produces less CO_2 than petrol. The LPG is stored under pressure as a liquid because in this state it is 250 times more dense that in its gaseous form. A standard petrol car can often be converted to run on LPG (this is known as **retrofitting**) by adding a second tank and fuel system for the LPG. Because the original petrol tank and petrol fuel operating system stay, the car becomes 'dual fuel'.

Hybrid Electric & Alternative Automotive Propulsion

Retrofitting – aftermarket vehicle conversion.

Benefits of using liquefied petroleum gas LPG include:

- smoother, quieter and cleaner running engines than those using conventional petrol
- the life span of the engine can be extended by as much as 30%
- lower emissions output, leading to less environmental pollution
- much lower fuel costs
- as the engine runs cleaner, servicing costs are reduced
- if the vehicle is designed to be a dual fuel car, (one that operates on either petrol or LPG) its range is increased
- resale values may be increased because the car is cheaper to run than its competitors
- the storage container for LPG in cars is normally of stronger construction than that of petrol tanks and as a result, crash damage safety is improved
- LPG has a high ignition temperature, around twice that of petrol, making it less likely to spontaneously combust and reducing the risk of fire

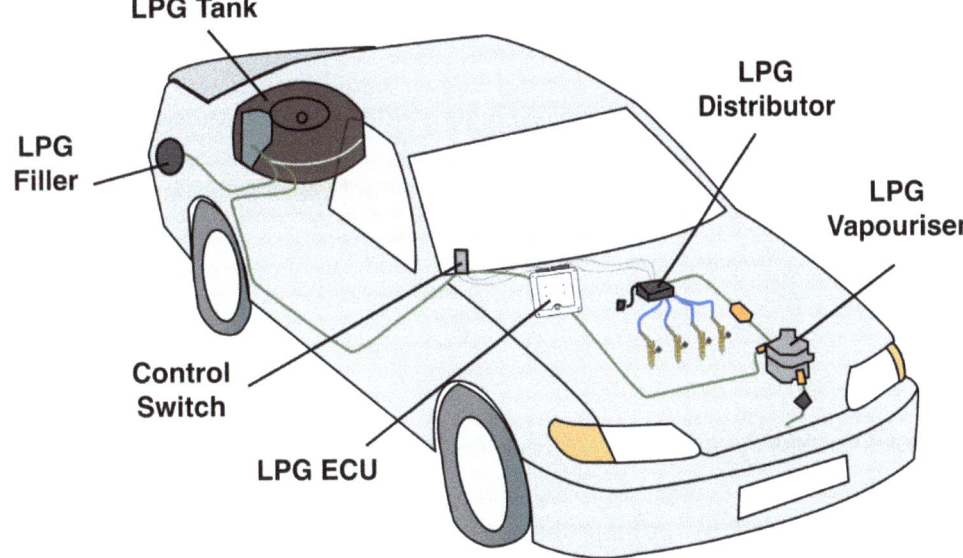

Figure 3.17 An LPG equipped car

Low Carbon Technologies

Remember that an LPG tank will never really be empty. As fuel is used and pressures fall, eventually the propane left over will remain as a gas.

An LPG tank should also never be completely filled. It is recommended that tanks are only taken to around 80% capacity to allow for expansion and contraction of the propane due to different ambient temperatures.

Unfortunately, LPG has a lower energy density than petrol and therefore fuel consumption is increased, but a lower duty of tax will often offset this higher consumption as a lower overall cost.

When compared to conventional petrol or Diesel engined vehicles, the emissions produced are far less harmful to health and the environment. Table 3.6 shows how petrol and Diesel compare to liquefied petroleum gas LPG.

Table 3.6 Emissions comparison between petrol, Diesel and LPG

LPG compared to petrol	LPG compared to Diesel
75% less carbon monoxide	60% less carbon monoxide
40% less oxides of nitrogen	90% less oxides of nitrogen
87% less potential of forming ground level ozone	70% less potential of forming ground level ozone
85% less hydrocarbons	90% less particulate matter (soot)
10% less carbon dioxide	

To combust properly inside an engine, petrol and Diesel need to be vaporised, as it is the fumes that burn not the liquid fuel. Complicated methods are needed to introduce petrol or Diesel, such as a carburettor or fuel injection, and turn it into a vapour. An advantage of using LPG as alternative fuel source to power vehicles is its very low boiling point of between - 0.6 degrees Celsius for butane and -42 degrees Celsius for propane. Because of its very low boiling point, most LPG used for vehicles is based on propane. With LPG you only require some form of simple nozzle to introduce the fuel and it will vaporise easily.

Converting a standard car to run on LPG

Although LPG is a very safe fuel to use for vehicle propulsion, care must be taken when converting a car to run on propane, otherwise safety may be compromised.

If the car is to be operated as a dual fuel vehicle, a second tank must be fitted to store the LPG. The tank may be cylindrical in shape and need a mounting space in the boot, or it can be round and take the place of the spare wheel. A filling hose will be required and an external filling point will need to be added to the vehicle body, normally close to the location of the petrol filling point. LPG from the tank can then be transferred to the engine via copper piping routed along the underside of the car.

A solenoid valve and filter are mounted in the fuel delivery line in order to remove dirt particles and prevent the flow of LPG when a dual fuel engine is being run on petrol.

A small regulator unit is then mounted on the engine's intake system, which warms the propane with heat from the engine cooling system, turning it into gas.

A mixer/distributor takes information from various engine sensors or ECU and introduces a controlled amount of gas to the intake manifold via injectors. The regulator and mixer unit will normally include a safety circuit which will cut the flow of propane gas to the inlet manifold if the engine should stall or cut out.

Special electronic circuitry is then required so that:

- a functioning fuel gauge can be created for the LPG system
- an automatic or manual switching system can be created to allow the engine to swap between petrol and LPG
- petrol fuel injection can be simulated electronically to the ECU to avoid the engine management system storing diagnostic trouble codes when the engine is running on LPG; this is achieved using a component known as an 'emulator'

Service an LPG system

How to:

1. Wearing appropriate PPE, conduct a visual inspection of all the LPG system components to check the security and condition.
2. Remove system pressure by closing the gas shut-off valve, starting the engine, and allowing it to stall. This will use up any liquid LPG in the supply pipes.
3. Locate the fuel filter in the supply line and carefully undo the pipes. Work in a well-ventilated area to ensure that any LPG gas leakage does not cause a fire risk.
4. Remove any fuel filter mountings and replace filter.
5. Once fitted, open, the gas valve and re-pressurise the system.
6. Check for leaks on the filter pipe work using a gas test leak solution.
7. Start and run the engine so that it reaches normal operating temperature and fully warms all system components; especially the LPG vaporiser. (Then switch off).
8. Locate the LPG vaporiser (also known as the reducer) in the engine compartment.
9. Place a container under the vaporiser, open the drain bung and allow any sludge build-up to fully drain from the system.
10. Close vaporiser drain bung.
11. Run engine and check for correct function and operation.

Compressed natural gas (CNG)

An alternative to LPG is high pressure compressed natural gas (CNG). CNG can be produced from a number of sources and is mainly composed of methane. When crude oil is extracted from the ground, around 3% is natural methane gas.

It can be used to fuel normal internal combustion engines instead of petrol. The combustion of methane produces the least amount of CO_2 of all fossil fuels. Petrol cars can be retrofitted to CNG and become bi-fuel natural gas vehicles (NGV) in a similar way to LPG conversion. Because the original petrol tank and petrol fuel operating system stay, the car becomes 'dual fuel'.

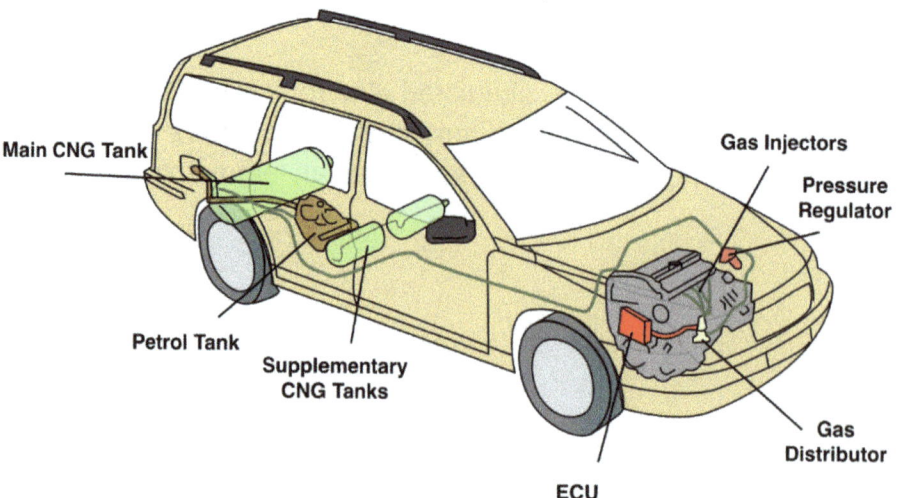

Figure 3.18 A CNG equipped car

The construction and operation of a CNG system is similar to that used in LPG, but instead of the fuel being stored in a liquid state, it is stored as a gas. If methane was to be stored as a liquid it would not only require high pressures, but also cooling to extremely low temperatures.

A disadvantage of this system compared to LPG is the density of the stored fuel. As a result, CNG vehicles will require a larger fuel tank in comparison to LPG in order to store similar quantities.

An advantage of this system however, is that compressed natural gas can be stored at a far lower pressure, in a form known as absorbed natural gas (ANG), and vehicles can be refuelled from the normal natural gas network without any further compression. This improves the logistics for making this fuel widely available to customers, as an infrastructure already exists which could be adapted for vehicle use.

Biogas

Biogas normally refers to a gas produced by the biological breakdown of **organic matter** in the absence of oxygen. Biogas normally comes from **biogenic** material, such as dead plant and animal matter, and is a type of renewable biofuel. A biogas generator plant will normally contain a unit called an anaerobic digester where plant material, animal matter and sewage are allowed to decompose and form methane, carbon dioxide and small amounts of hydrogen sulphide. After purification of the raw gas, compressed biogas can be used to power normal internal combustion engines, in a similar manner to compressed natural gas (CNG).

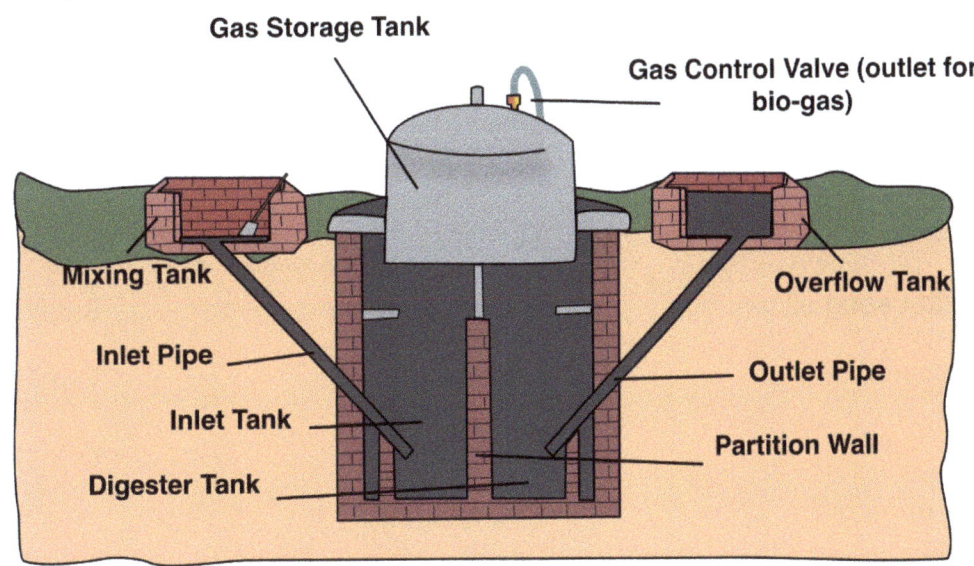

Figure 3.19 Bio-gas production

Anaerobic - without oxygen.

Organic matter – something that has come from a once living organism and is capable of decay.

Biogenic – produced or brought about by living organisms.

Biodiesel

Biodiesel (fatty acid methyl ester) is a way of making a form of Diesel fuel from a very wide range of oil-producing plants, such as:

Algae oil	Jojoba oil	Radish oil
Artichoke oil	Karanj oil	Rapeseed oil
Canola oil	Kukui nut oil	Rice bran oil
Castor oil	Milk bush	Safflower oil
Coconut oil	Pencil bush oil	Sesame oil
Corn	Mustard oil	Soybean oil
Cottonseed oil	Neem oil	Sunflower oil
Flaxseed oil	Olive oil	Tung oil
Hemp oil	Palm oil	Waste vegetable oil (WVO)
Jatropha oil	Peanut oil	

Many of these oils can be commercially refined and sold as an alternative to standard Diesel.

Waste vegetable oil

Waste vegetable oil can often be used as a direct substitute for normal Diesel, although it can cause engine and fuel system problems. If the oil is unfiltered, the relatively high **viscosity** and any debris, may cause damage to the fuel pumps and block the fuel filter. The viscosity also leads to poor atomisation of the fuel at the injectors and inefficient combustion. Poor combustion leads to high carbon deposits inside the cylinder and around the injectors and valves. Incomplete combustion may also cause contamination of the engines lubricating oil. The best way to prevent these issues is to refine the oil through a process known as transesterification.

Once treated the waste vegetable oil will have:

- A lower viscosity
- A lower boiling point
- A lower flash point
- Damaging glycerides removed

Viscosity - a fluids resistance to flow.

Transesterification - the process of exchanging the alcohol group of an ester compound with another alcohol.

Transesterification

Oils produced from plants or animal fat will need fully refining through a process known as transesterification before they can be used as a biodiesel. This procedure involves the separation of ester from the glycerides found in these oils.

To do this, the fat/oil is reacted with alcohol using a chemical catalyst such as potassium or sodium hydroxide. The mixture is often heated in a sealed container which is kept just above the boiling point of the alcohol. The heavier glycerol settles out and can then be removed, recycled and sold on to the pharmaceutical industry.

In order for the separated ester to be used as fuel, the alcohol must then be removed, which is done by a process of flash evaporation or distillation. When complete the biodiesel is then checked to ensure that:

- All of the glycerine has been removed
- All of the alcohol has been removed
- All of the catalyst chemical has been removed
- The fatty acids have been removed

Once this is done, the resulting biodiesel should act as an adequate fuel for use in internal combustion engines.

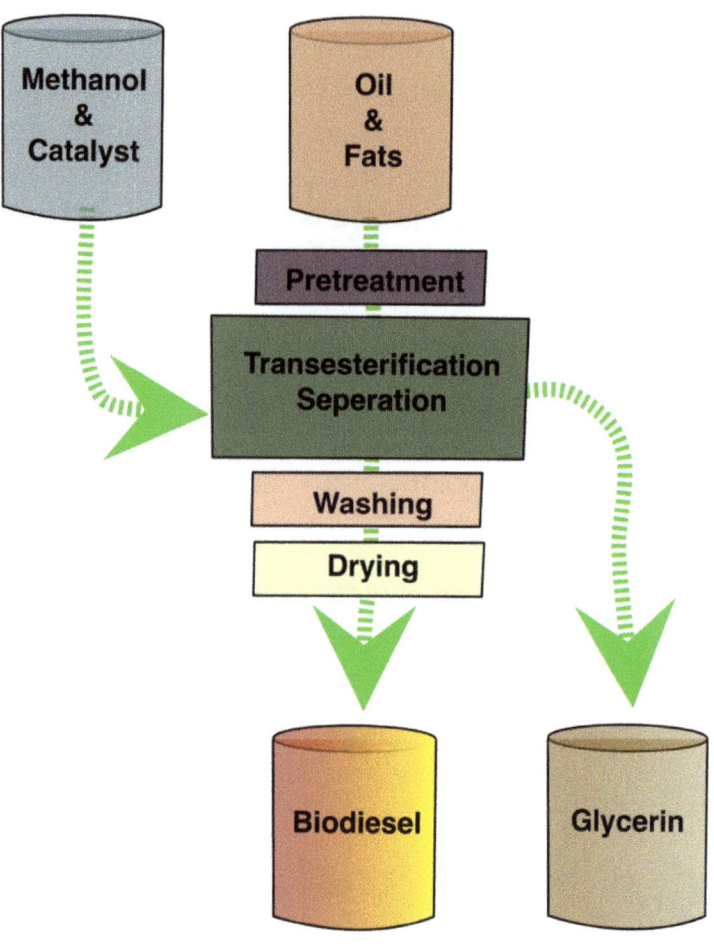

Figure 3.20 Transesterification

Table 3.7 shows some of the advantages and disadvantages of biodiesel when compared to fossil Diesel.

Table 3.7 The advantages and disadvantages of biodiesel

Advantages of biodiesel	Disadvantages of biodiesel
Because biodiesel is mainly produced from plants or crops, it is considered a renewable energy source when compared to fossil fuels.	Biodiesel has a lower energy density than fossil Diesel fuel. This means that biodiesel vehicles are not quite able to keep up with the fuel economy of a normal fossil-fuelled Diesel vehicle.
Biodiesel is often considered carbon neutral when used as a vehicle fuel. This is because the carbon dioxide released during combustion is equal to the carbon dioxide absorbed from the atmosphere during the plants life span.	The production of biodiesel can be expensive and use a large amount of energy. The energy needed often comes from non-renewable sources and therefore creates carbon dioxide from production.
Biodiesel is completely biodegradable and non-toxic, as a result fuel spillages are less of a risk to environmental pollution than fossil fuels.	Some biodiesel, especially those produced from waste vegetable oil, contain chemicals which can damage natural rubber leading to leaks from fuel system seals.
Biodiesel has a higher **flash point** than fossil diesel, reducing the risk of fire in the event of an accident.	Large areas of land are required to grow the oil producing crops. This often leads to reduced area for the production of food crops or deforestation.

Flash point – the temperature at which a fuel forms an ignitable mixture in air.

Biodiesel from algae

Biodiesel can also be created from algae. Algae can be grown in open ponds or sealed systems known as 'photo bioreactors'. Algae is one of the most efficient organisms on earth, due to its very rapid growth rate and can be harvested in areas which do not affect the production of food crops. Once dried and pressed to remove the oil, algae produces around 300 times more oil than standard biofuel crops and can be harvested on a 1 to 10 day cycle rather than yearly.

Replace the fuel filter on a biodiesel system

How to:

1. Wearing the appropriate PPE, raise the vehicle off the ground.
2. Locate the fuel filter near the fuel tank and position the drainer under the filter.
3. Remove the filter from the mounting using the socket and wrench.
4. Wipe around the filter pipes with a rag to prevent dirt getting into the pipes. Remove the filter pipes with the pipe release tool.
5. Refit the new filter in the reverse order of removal and bleed the system if required.
6. Lower the vehicle.
7. Run the engine and check for leaks.

Bioalcohol/Ethanol

Different types of alcohol are often able to be used as a fuel source for internal combustion engines. The two main types of alcohol used are methanol and ethanol. Methanol is often created from natural gas and is therefore not normally considered a biofuel (although it can be produced from biogas, but this is often a more complex and expensive procedure).

Ethanol is mainly produced from biological material through a process of fermentation, which is similar to that used in the production of alcohol for drinking. Ethanol can also be produced from petroleum based products, but would be very dangerous to drink; causing blindness or death.

The alcohols used for fuel have a lower energy density than conventional petrol or Diesel, with around one and a half, to two times the volume required to produce the same amount of heat energy. However, alcohol has a naturally high **octane**, meaning that it can be run at higher compression pressures inside the engine which then produces similar results in performance and fuel economy. There is also a reduction in tail pipe emissions when compared to petrol or Diesel.

A disadvantage of alcohol is that it tends to be corrosive or promote corrosion in fuel system components. Over a period of time it is possible that fuel systems may become excessively corroded or blocked. In recent years many manufactures have been designing cars with fuel systems that are able to tolerate around 10% ethanol which is often now blended with standard petrol, but to run solely on alcohol, fuel system materials need to be adapted and engine management systems reprogrammed to take into account its use.

Diesel engines are also able to operate on alcohol, but as the fuel has a low **cetane** value, additives such as glycol need to be combined with the fuel to improve ignition.

Hybrid Electric & Alternative Automotive Propulsion

Octane - a colourless flammable hydrocarbon with the ability to suppress detonation.

Cetane - a measure of a fuels ignition delay (the time period from injection until it starts to burn).

Other alcohols that can be used as a fuel include butanol and propanol. These are more expensive and difficult to produce than methanol and ethanol making them less viable for fuel production. If a method can be found to produce butanol economically, then fuel economy and performance could be increased, as it has a very similar energy density to petrol.

The first car that you could buy that used ethanol as a fuel was the Model T Ford, produced from 1908 until 1927. It was fitted with a carburettor which had adjustable jetting, allowing use of petrol or ethanol, or a combination of the two.

Ammonia Green

Ammonia Green (NH_3) is being used with success by some vehicle manufacturers because it can run in spark ignition or compression ignition engines with only minor modifications.

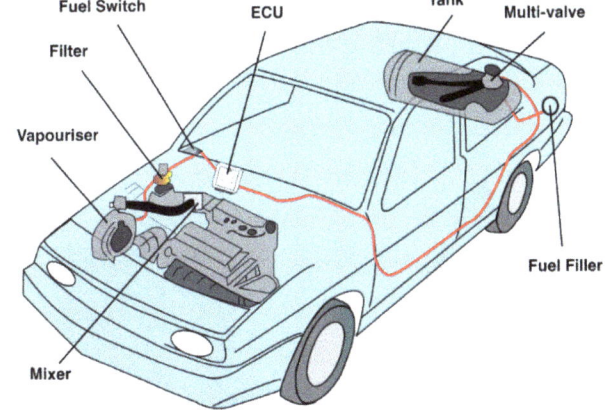

Figure 3.21 An ammonia green (NH_3) equipped vehicle

Using ammonia to power internal combustion engines, provides some advantages over other fuel sources:

- It has a very high energy density when compared to other non-petroleum based fuels.

- It contains no carbon, and therefore releases no carbon dioxide during combustion and generates no particulate matter.

- As ammonia can be manufactured relatively easily, it won't run out like fossil fuels.

Ammonia is manufactured by reacting hydrogen with the nitrogen in air to produce NH_3, meaning the raw materials needed to produce ammonia are water and air.

The most common method of creating hydrogen for the production of ammonia comes from natural gas in a process which can cause a significant amount of greenhouse gasses to be released to atmosphere. This is known as brown ammonia.

If the hydrogen for the production of ammonia is created through electrolysis, then green non-polluting methods can be used, these include:

- Wind power
- Solar
- Hydroelectric
- Wave power

If the ammonia is produced through electrolysis from renewable energy sources, it is known as ammonia green. Liquid ammonia has half the density of petrol or Diesel, this means that it can be easily carried in sufficient quantities in vehicles, and unlike some other alternative fuels and produce a good driving range. On combustion it produces no emissions other than nitrogen and water vapour, making the exhaust non-polluting.

Ammonia is considered to be very toxic, but if stored and handled correctly, it is no more dangerous than petrol or LPG.

- What gas is CNG mainly composed of?
- List six oil producing plants that can be used to produce biodiesel.
- What is brown ammonia?

Electricity

Electric motors are a very effective method of providing a source of **propulsion** for cars and they produce no emissions while in use. Unfortunately, many of the methods used to create the electricity needed to drive these motors are not very **efficient** or can be polluting to the environment. Currently, renewable sources of electricity such as wind, solar and hydro-electricity are unable to meet the supply demands needed to make electric powered vehicles truly non-polluting.

Environmentally friendly methods used to create electricity for powering electric vehicles mainly comes from:

- Solar power
- Mains supply
- Hydrogen fuel cells
- Hybrid drive.

For more information on hybrid and electric vehicles, see Chapter 2.

Propulsion – the action of driving or pushing forward.

Efficient – how well something works.

Solar

A solar car is an electric vehicle powered by energy from the sun, which is obtained from solar panels on the car. A solar panel converts light energy into electricity that can be used as a source of power. The sun gives off approximately 1000 watts of energy for every square metre of the earth's surface. Solar panels that are sometimes found on cars, are also known as photovoltaic cells. This comes from the words "photo" meaning light and "voltaic" meaning electricity. Photovoltaic cells, are made of **semiconductor** material such as silicon. When sunlight strikes the semiconductor material, some of the energy is absorbed and is converted into moving electrons within the material to create an electric current. The flow of electrons, is in one direction (direct current). If electrical contacts are placed above and below the photovoltaic cell, the electricity produced by converting photons can be used to power electrical circuits.

Semiconductor - a material with the ability to switch its properties between an insulator and a conductor.

The silicone used in a photovoltaic cell works in this way because its atoms are not completely filled up with electrons in their outer orbit or shell. If small amounts of impurity are added to the material, such as phosphorus, small amounts of extra electrons are available in the material making it negative. On the other hand, if boron is added to the silicone, less electrons are available, making it positive.

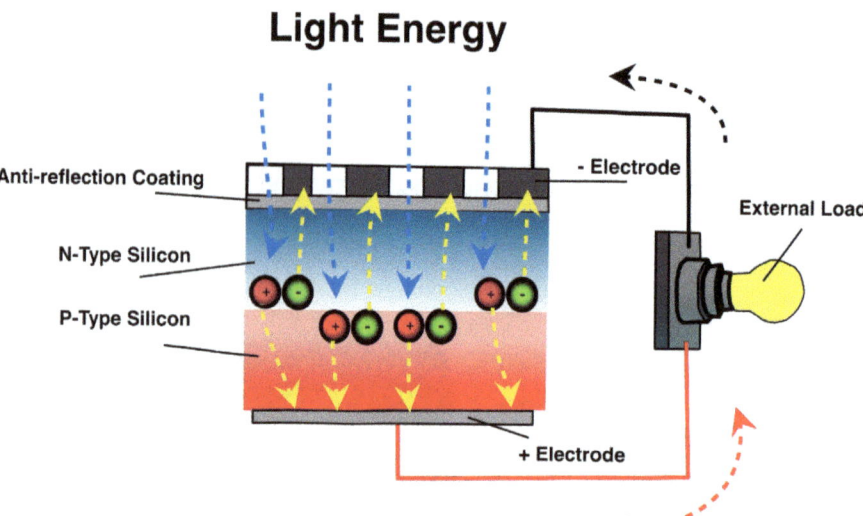

Figure 3.22 A photovoltaic cell

These two sections of silicone can then be joined, and if supplied with light from the sun, share the electrons between the positive and negative sections creating electric current.

Unfortunately not all of the energy provided by the sun can effectively be converted into electricity. This is because light comes in different wavelengths and photovoltaic cells can only use certain areas of the electromagnetic spectrum. Currently solar panels cannot be used to directly supply a car with enough power to provide drive to electric motors, but they can be used to charge batteries or extend the range of 'plug in' electric vehicles.

Mains supply

A number of manufacturers are producing a range of mains electricity charged electric cars. Instead of an internal combustion engine, these vehicles (often known as 'plug in') are powered from high capacity batteries that drive electric motors. Although these vehicles produce no emissions when they are driven, mains generated electricity is often created using fossil fuel (which creates pollution) or atomic energy (which is dangerous and radioactive). The main limitations of 'plug in' electric vehicles are the distance they can travel on a single charge (known as range), and the amount of time it takes to recharge the batteries, which can be many hours. For more information on electric vehicles, see Chapter 2.

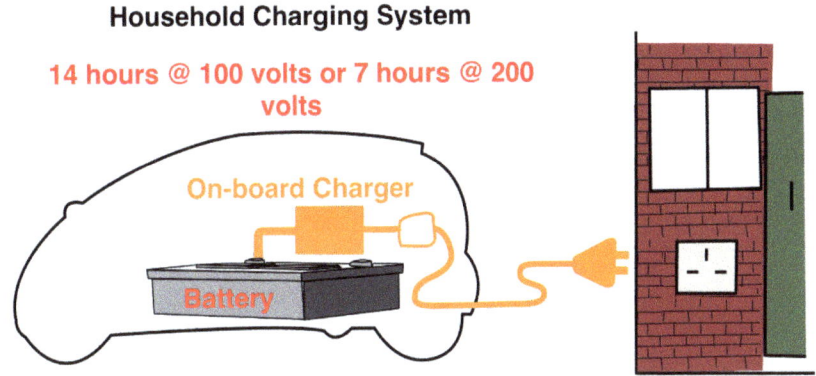

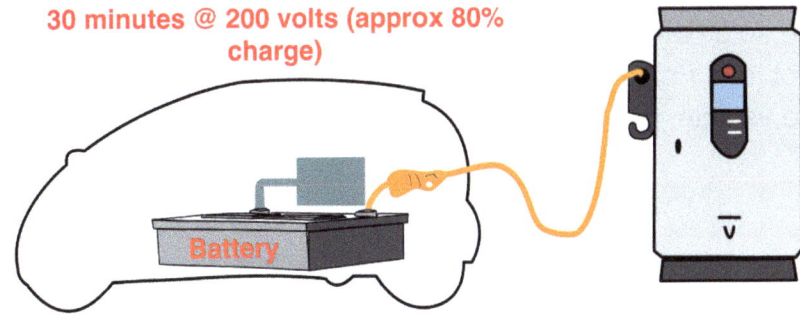

Figure 3.33 A plug-in electric car

Hydrogen

Hydrogen is the most abundant gas in the universe and it makes an extremely good fuel for operating vehicles. Hydrogen is extremely flammable and produces very little in the way of harmful emissions (the main by-product being water). Unfortunately, hydrogen does not occur naturally on earth so must be manufactured.

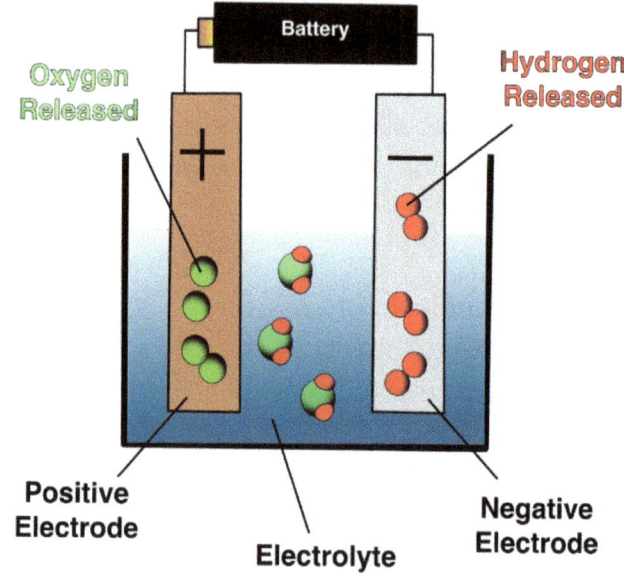

Figure 3.34 Making hydrogen by electrolysis

The process of making hydrogen is fairly straightforward. It normally involves separating the hydrogen and oxygen in water by a process of **electrolysis**. However, in general terms, it takes about three times as much energy to make the hydrogen as can be obtained from the hydrogen itself. This makes hydrogen an inefficient fuel source in many ways. Also, hydrogen **molecules** are so small that they will leak through almost any container. This means that storage can be a problem. For example, if you filled up a standard fuel tank with hydrogen, even if you didn't use the vehicle, the tank would be empty in a few days.

Electrolysis - a chemical decomposition produced by passing an electric current through a liquid.

Molecules - the smallest component of a chemical element.

Hydrogen internal combustion engines (HICE)

The gas hydrogen can be used to directly power internal combustion engines with a few modifications. The advantages of running an engine on hydrogen is that the only emission caused by combustion should be water vapour, and the fuel has the potential to provide a higher power output than petrol or Diesel. Producing an engine that runs on hydrogen is not difficult, but getting an engine that runs well on hydrogen is a different matter.

To allow an engine to run on hydrogen will normally require the adaption of the fuel delivery system. Hydrogen can be introduced with three main methods:

- Single-point injection
- Multi-point injection
- Direct injection

Single-point and multi-point require the least amount of engine adaption and hydrogen becomes a gas easily inside the intake manifold. Direct injection may need the engine to be considerably modified but does offer the best range of operation when running on hydrogen as the air fuel mixture can adjusted in far greater detail.

The ignition system on a hydrogen internal combustion engine can be similar to a standard petrol ignition system, although cold running spark plugs will be required to prevent **pre-ignition**. Also platinum electrode spark plugs cannot be used as this metal acts as a catalyst and reacts with hydrogen causing it to oxidise with air.

Pre-ignition - a condition where the fuel in the combustion chamber ignites before the spark has been produced.

Although hydrogen is a good source of fuel for vehicles, its properties have advantages and disadvantages when used in internal combustion engines.

Table 3.8 describes some advantages and disadvantages of using hydrogen in internal combustion engines.

Table 3.8 The advantages and disadvantages of hydrogen when used in HICE

Property	Advantages	Disadvantages
Its wide range of flammability.	Hydrogen will burn when used with extremely weak mixtures that are well below the recommended stoichiometric values. This has the advantage of easy engine starting and a more complete combustion overall.	There is a limit to how weak an air fuel mixture can become before the combustion temperature falls to a point where power output from the engine is reduced.
Its low ignition energy requirements.	The amount of energy needed to ignite hydrogen is far less than that needed for petrol. This means that a standard spark ignition system can be used to ignite the fuel, even with extremely weak mixtures.	Because of hydrogens low ignition energy requirements, hot spots inside the cylinder may cause pre-ignition, leading to a misfire. Sources of combustion cylinder hot spots could be carbon buildup from burnt lubrication oil, or overheated spark plug electrodes.
Its ability to resist quenching.	Unlike petrol which will sometimes be extinguished during its combustion as it approaches the cooler cylinder walls, hydrogen is less likely to be quenched and therefore use up more of its available energy.	The small quenching distance can increase the risk of a backfire as the combusting hydrogen is able to burn closer to a nearly closed inlet valve. If the induction system uses single-point or multi-point injection in the manifold, it is possible that fuel here will ignite and cause a fire.
Its high auto ignition temperature.	A high auto ignition temperature is important to the stability of the fuel, meaning that it resists detonation solely due to air temperature. This stability means that it can run at far higher compression ratios than a standard petrol engine and as compression ratios rise, so does the performance output.	The high auto ignition temperature makes hydrogen unsuitable for use in compression ignition engines.

Table 3.8 The advantages and disadvantages of hydrogen when used in HICE

Its ability to spread out due to its low density.	As hydrogen is a gas even at very low temperatures. If a leak in the system occurs, it will spread out and disperse very quickly reducing the risk of an accidental fire.	The low density of hydrogen means that it takes up a lot more space in an engine cylinder than vaporised petrol. This means that the energy density available for combustion is much lower, reducing the engines performance output.
		Another issue caused by low density is that a very large volume of hydrogen needs to be stored to give the vehicle an adequate driving range.

Notes: The ideal are fuel ratios for a petrol engines are 14.7:1 (by mass) compared to 34:1 (by mass) for hydrogen. Although this is the recommended air fuel ratio, hydrogen internal combustion engines will often run far weaker than this, down to values as low as 180:1.

Hydrogen fuel storage

Unlike petrol and Diesel, which attach their hydrogen atoms to carbon so the fuel can be stored in a liquid form, hydrogen is a gas at normal temperatures. This means that to store pure hydrogen on board a vehicle as a liquid would require extremely high pressures and low temperatures.

Crankcase ventilation

As with a normal internal combustion engine, blow-by of combustion mixtures from a hydrogen internal combustion engine can leak past the piston rings and into the crankcase. Crankcase ventilation needs to be correctly controlled for two main reasons.

1. A build-up of hydrogen in the crankcase could lead to an explosion, igniting other flammable materials such as oil.

2. A by-product of hydrogen combustion with air is water. Water can contaminate the engine oil, reducing its lubrication properties.

Figure 3.35 Water contamination of the engine oil

Hydrogen fuel cell

Another method that can be used to power vehicles using hydrogen, other than burning it inside an engine, is a fuel cell. Some manufacturers are now producing cars with a hydrogen fuel cell, but because of their complexity, they are currently expensive. The fuel cell uses hydrogen to create electricity that can then be used to power electric motors.

A standard vehicle battery, including those found in hybrid and electric vehicles, stores all of its chemical energy inside a casing which it uses to create electricity through a chemical reaction. Once the battery has used up all of its chemical energy, the battery is flat. Many car batteries can reverse this process by supplying an electric current with a voltage potential higher than that coming out of the battery. This is normally achieved using an alternator or generator. A fuel cell is similar to a battery but it doesn't store its own internal electricity in the form of a charge. With a fuel cell, as long as the cell is kept supplied with fuel, in this case hydrogen and oxygen, then it works like a battery that never goes flat. Hydrogen is stored in a separate container/fuel tank, and then mixed with the oxygen inside fuel cell to create electricity.

The hydrogen fuel cell is not a new idea. Sir William Grove invented the first fuel cell in 1839. He knew that water could be separated into hydrogen and oxygen if electric current is passed through it by a process known as electrolysis. Grove found that if the process was reversed, he could create electricity by recombining the hydrogen and oxygen and creating water. He then went on to create a very basic type of fuel cell which he called a gas voltaic battery.

Fuel cell construction

The most common type of hydrogen fuel cell is made using a component called a **proton** exchange **membrane** (PEM). This is a material that separates the two sides of the fuel cell.

- One side of the fuel cell is fed with oxygen from the surrounding air.
- The other side is fed with hydrogen from a fuel tank.

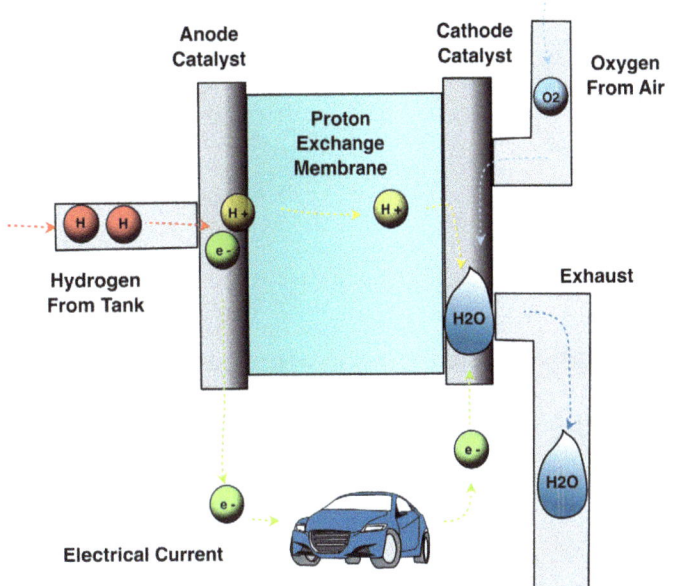

Fuel cell operation

As hydrogen enters the fuel cell, a reaction takes place that strips the protons from the hydrogen atom, and moves them through the membrane towards the oxygen on the other side. This leaves the **electrons** from the hydrogen atoms, which travel through a different circuit and create electric current. After the energy from the hydrogen has been converted into current and powered the vehicles electric circuit, the electrons reattach themselves to the protons and the hydrogen atom combines with the oxygen to form H_2O. This means that the only emission from the fuel cell is water, making it clean and non-polluting.

Figure 3.36 Hydrogen fuel cell

The typical output from a single fuel cell is approximately 0.8V. This means that a number of fuel cells have to be combined (known as a fuel cell stack) to create a useable amount of voltage to drive electric motors.

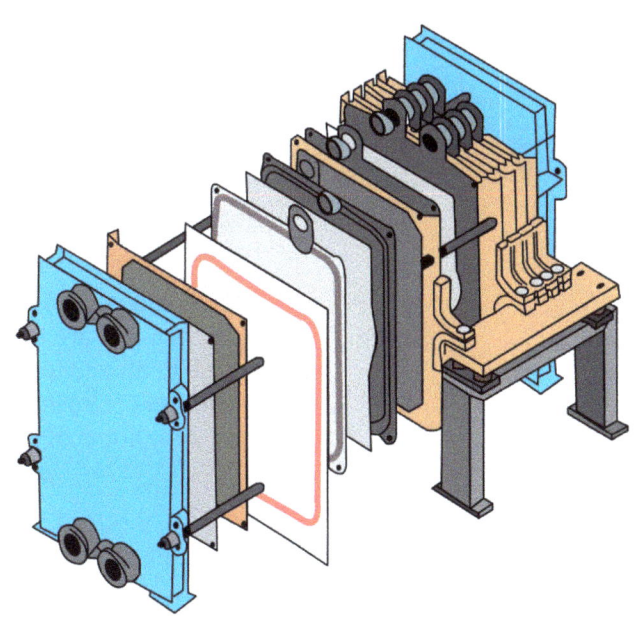

Figure 3.37 A fuel cell stack

Protons – the positively charged particles of an atom.

Membrane – a thin layer of material that is used to separate two connected areas.

Electrons – the negatively charged particles of an atom.

Alternative propulsion comparisons

Table 3.9 shows comparisons in efficiency for different fuel and vehicle drive types.

Table 3.9 System drive types and efficiency

Fuel/drive system type	Efficiency
Steam produced by coal	In practice, a steam engine exhausting the steam to atmosphere will typically have an efficiency (including the boiler) in the range of 1-10%; this means that it is around 90% inefficient. With the addition of a condenser and multiple expansion, this figure can be improved to 25% or better.
Petrol	The efficiency of a petrol-powered car is surprisingly low. All of the heat that comes out as exhaust or goes into the radiator is wasted energy. The engine also uses a lot of energy turning the various pumps, fans and generators that keep it going. So the overall efficiency of a petrol engine is about 20%. That is, only about 20% of the thermal-energy content of the petrol is converted into mechanical work. With the introduction of gasoline direct injection petrol engines (GDI), manufacturers are beginning to see improvements in efficiency of the petrol engine to around 35%. This is because the amount of fuel used is more accurately controlled and stop start systems have been introduced which take away some of the wasteful energy losses that occur at idle.

Table 3.9 System drive types and efficiency

Diesel	Diesels, are able to reach an efficiency of about 40% in the engine speed range of idle to about 1800 rpm. After this speed, efficiency begins to decline due to air pumping losses within the engine. Modern turbo-Diesel engines are using electronically controlled, common-rail fuel injection, that increases the efficiency up to 50% with the help of variable vane turbo-charging systems; this also increases the engines' torque at low engine speeds (1200-1800RPM).
Hybrid	Depending on the hybrid drive type: Series Parallel Combination Efficiency values will vary. If you combine the efficiency of a petrol engine of around 20%, with the efficiency of electric motors of around 90%, then in theory you should get an overall efficiency of about 70%. Unfortunately, due to construction and design, heat losses will reduce this figure to around 55% to 60%, putting them around the same sort of efficiency as a modern common rail Diesel. These figures though do not take into account the large reduction in emissions produced by hybrids when compared to Diesel engines.
Plug in electric	A battery-powered electric car has a fairly high efficiency. The battery is about 90% efficient (most batteries generate some heat, or require heating), and the electric motor/inverter is about 80% efficient. This gives an overall efficiency of about 72%. But that is not the whole story. The electricity used to power the car had to be generated somewhere. If it was generated at a power plant that used a combustion process (rather than nuclear, hydroelectric, solar or wind), then only about 40 percent of the fuel required by the power plant was converted into electricity. The process of charging the car requires the conversion of alternating current (AC) power to direct current (DC) power. This process has an efficiency of about 90 percent. So, if we look at the whole cycle, the efficiency of an electric car is 72% for the car, 40% for the power plant and 90% for charging the car. That gives an overall efficiency of 26%. The overall efficiency varies considerably depending on what sort of power plant is used. If the electricity for the car is generated by a hydroelectric plant for instance, then it is basically free (no fuel was burned to generate the electricity), and the efficiency of the electric car is about 65%.

Table 3.9 System drive types and efficiency

Solar	Because photovoltaic cells (solar panels) are unable to convert all of the energy in the electromagnetic spectrum produced by sunlight, they are only currently around 10% efficient. This means that a 1 square meter solar panel that is capable of receiving 1 kilo watt of energy from the sun can only convert this into 100 watts of energy. This is around 1.3 horse power (UK). This very low power output makes solar panels unsuitable for powering cars but they can be used to extend the range of plug-in electric vehicles or be used in the electrolysis processes used to create hydrogen or ammonia which can then be used in vehicle propulsion.
Hydrogen powered ICE	When compared to a standard internal combustion engine running on petrol, a hydrogen engine has the potential to produce a greater power output because of the higher compression ratios that can be used. Unfortunately, due to the space required inside the engine cylinder by the volume of hydrogen in its gaseous state, air/fuel ratios tend to run extremely weak leading to a low efficiency overall of around 25%.
Hydrogen fuel cell	If the fuel cell is powered with pure hydrogen, it has the potential to be up to 80% efficient. That is, it converts 80% of the energy content of the hydrogen into electrical energy. However, we still need to convert the electrical energy into mechanical work. This is accomplished by the electric motor and inverter. A reasonable number for the efficiency of the motor/inverter is about 80%. So we have 80% efficiency in generating electricity, and 80% efficiency converting it to mechanical power. That gives an overall efficiency of about 64%.

- What type of semiconductor material is used to construct a photovoltaic cell?
- State two advantages and two disadvantages of a HICE engine.
- In relation to fuel cells, what do the letters 'PEM' stand for?

Common Acronyms/Abbreviations

A - Amperes

A/C - Air Conditioning

A/F - Air Fuel Ratio

A/T - Automatic Transmission

AAC - Auxiliary Air Control Valve

AAT - Ambient Air Temperature

ABD - Automatic Brake Differential

ABS - Antilock Brake System

ABV - Air Bypass Valve

AC - Alternating Current

ACC - Automatic Climate Control

ACC - Air Conditioning Clutch

ACR - Air Conditioning Relay

ACR4 - Air Conditioning Refrigerant, Recovery, Recycling, Recharging

ACV - Air Control Valve

ADU - Analogue-Digital Unit

AEV - All Electric Vehicle

AFC - Air Flow Control

AFL - Advanced Front Lighting System

AFM - Air Flow Meter

AFR - Air Fuel Ratio

AFS - Air Flow Sensor

AGM - Absorbed Glass Matt

Ah - Amp Hours

AIR - Secondary Air Injection System

AIS - Automatic Idle Speed

ALC - Automatic Level Control

AM - Amplitude Modulation

API - American Petroleum Institute

APS - Atmospheric Pressure Sensor

ARC - Automatic Ride Control

ARS - Automatic Restraint System

ASARC - Air Suspension Automatic Ride Control

ATC - Automatic Temperature Control

ATDC - After Top Dead Centre

ATF - Automatic Transmission Fluid

ATS - Air Temperature Sensor

AVO - Amps Volts Ohms

AWD - All Wheel Drive

AWG - American Wire Gauge

AYC - Active Yaw Control

B/MAP - Barometric/Manifold Absolute Pressure

BARO - Barometric Pressure

BCM - Body Control Module

BCM- Battery Control Module

BDC- Bottom Dead Centre

BEV- Battery Electric Vehicle

BHP - Brake Horsepower

BOB - Breakout Box

BP - Barometric Pressure

BPP - Brake Pedal Position Switch

BTDC - Before Top Dead Centre

BTS - Battery Temperature Sensor

Btu - British Thermal Unit

BUS N - Bus Negative

BUS P - Bus Positive

C - Celsius

CA- Cranking Amps

CAN - Controller Area Network

CANP - **EVAP** Canister Purge Solenoid

CAS - Crank Angle Sensor

CBW- Clutch By Wire

CC - Catalytic Converter

CC - Climate Control

CC - Cruise Control

CC - Cubic Centimetres

CCA- Cold Cranking Amps

CD- Compact Disc

CDI - Capacitor Discharge Ignition

CFC - Chlorofluorocarbons

CFI - Continuous Fuel Injection

CI- Compression Ignition

CKP - Crankshaft Position Sensor

CL - Closed Loop

CLC - Converter Lockup Clutch

CLV - Calculated Load Value

CMP - Camshaft Position Sensor

CNG- Compressed Natural Gas

CO - Carbon Monoxide

CO2 - Carbon Dioxide

COC - Conventional Oxidation Catalyst

COP - Coil On Plug Electronic Ignition

COSHH- Control Of Substances Hazardous to Health

CP - Crankshaft Position Sensor

CP - Canister Purge (GM)

CPP - Clutch Pedal Position

CPU - Central Processing Unit

CRC - Cyclic Redundancy Check

CRD- Common Rail Diesel

CRS - Common Rail System

CTP - Closed Throttle Position

CTS - Coolant Temperature Sensor

CV - Constant Velocity

CVT - Continuously Variable Transmission

DBW - Drive By Wire

DC - Duty Cycle

DC - Direct Current

DCS - Dual Clutch System

DI - Distributor Ignition (System)

DI - Direct Ignition

DIS - Direct Ignition (Waste Spark)

DIS - Distributorless Ignition System

DMF - Dual Mass Flywheel

DMM - Digital Multimeter

DLC - Data Link Connector (OBD)

DOHC - Dual Overhead Cam

DPF - Diesel Particulate Filter

DRL - Daytime Running Lights

DTC - Diagnostic Trouble Code

DVD - Digitally Versatile Disc

EAIR - Electronic Secondary Air Injection

EBCM - Electronic Brake Control Module

EBP - Exhaust Back Pressure

EBD - Electronic Brake Force Distribution

ECC - Electronic Climate Control

ECM - Engine/Electronic Control Module

ECS - Emission Control System

ECT - Engine Coolant Temperature

ECU - Electronic Control Unit

EDC - Electronic Diesel Control

EECS - Evaporative Emission Control System

EEGR - Electronic EGR (Solenoid)

EEPROM - Electronically Erasable Programmable Read Only Memory

EFI - Electronic Fuel Injection

EFT - Engine Fuel Temperature

EGO - Exhaust Gas Oxygen Sensor

EGR - Exhaust Gas Recirculation

EGRT - Exhaust Gas Recirculation Temperature

EMF - Electromotive Force (voltage)

EMI - Electromagnetic Interference

EOBD - European On Board Diagnostics

EOP - Engine Oil Pressure

EOT - Engine Oil Temperature

EPA - Environmental Protection Act

EPB - Electronic Parking Brake

EPROM - Erasable Programmable Read

Only Memory

EPS- Electronic Power Assisted Steering

ESP- Electronic Stability Programme

ESS - Engine Start-Stop

EVAP - Evaporative Emissions System

EVAP CP - Evaporative Canister Purge

FM- Frequency Modulation

FOT - Fixed Orifice Tube

FSD- Full Scale Deflection

FT - Fuel Trim

FWD - Front Wheel Drive

GDI - Gasoline Direct Injection

GND - Electrical Ground Connection

GPS- Global Positioning System

GWP - Global Warming Potential

H – Hydrogen

HASAWA- Health and Safety at Work Act

H2O - Water

HC - Hydrocarbons

HCA- Hot Cranking Amps

HDI- High Pressure Direct Injection

HEGO - Heated Exhaust Gas Oxygen Sensor

HFC- Hydrogen Fuel Cell

HFC- Hydro-fluoro Carbon

HFO- Hydro-fluoro Olefin

Hg – Mercury

HICE- Hydrogen Internal Combustion Engine

HID- High Intensity Discharge (lighting)

HO2S - Heated Oxygen Sensor

hp - Horsepower

HSE- Health and Safety Executive

HT - High Tension

HUD - Head Up Display

HVAC - Heating Ventilation and Air Conditioning

Hz - Hertz

I/O - Input / Output

IA - Intake Air

IAC - Idle Air Control (motor or solenoid)

IAT - Intake Air Temperature

IC - Integrated Circuit

IC - Ignition Control

ICE- In Car Entertainment

ICE – Internal Combustion Engine

ICM - Ignition Control Module

IFS - Inertia Fuel Switch

IGBT- Insulated Gate Bipolar Transistor

IGN - Ignition

IGN ADV - Ignition Advance

IGN GND - Ignition Ground

IPR - Injector Pressure Regulator

ISC - Idle Speed Control

ISO - International Standard of Organisation

KAM - Keep Alive Memory

Kg/cm2 - Kilograms/ Cubic Centimetres

KHz - Kilohertz

Km - Kilometers

KPA - Kilopascal

KPI - Kingpin Inclination

KS - Knock Sensor

KWP - Keyword Protocol

l - Litres

LCD - Liquid Crystal Display

LED - Light Emitting Diode

LHD - Left Hand Drive

Li-ion - Lithium ion

LOOP - Engine Operating Loop Status

LOS - Limited Operating Strategy

LPG - Liquefied Petroleum Gas

LSD - Limited Slip Differential

LTFT - Long Term Fuel Trim

LWB - Long Wheel Base

M/T - Manual Transmission

MAC - Mobile Air Conditioning

MAF - Mass Air Flow Sensor

MAP - Manifold Absolute Pressure Sensor

MAT - Manifold Air Temperature

MCM - Motor Control Module

MEF - Methane Equivalency Factor

MF - Maintenance Free

MFI - Multiport Fuel Injection

MIL - Malfunction Indicator Lamp

MPG - Miles Per Gallon

MPH - Miles Per Hour

mS or ms - Millisecond

mV or mv - Milivolt

N - Nitrogen

NCAPS - Non-Contact Angular Position Sensor

NCRPS - Non-Contact Rotary Position Sensor

NGV - Natural Gas Vehicles

Ni-MH - Nickel Metal Hydride

Nm - Newton Meters

NOx - Oxides of Nitrogen

NPN - Negative Positive Negative

NTC - Negative Temperature Coefficient

O2 - Oxygen

OBD I - On Board Diagnostics Version I

OBD II - On Board Diagnostics Version II

OC - Oxidation Catalytic Converter

OD - Overdrive

OD - Outside Diameter

ODP - Ozone Depletion Potential

OE - Original Equipment

OEM - Original Equipment Manufacturer

OFN - Oxygen Free Nitrogen

OHC - Overhead Cam Engine

OHV - Overhead Valve

OL - Open Loop

OS - Oxygen Sensor

P/N - Part Number

PAG - Polyalkylene Glycol

PAIR - Pulsed Secondary Air Injection

PATS - Passive Anti-Theft System

PCB - Printed Circuit Board

PCM - Powertrain Control Module

PCV - Positive Crankcase Ventilation

Pd -Potential Difference (volts)

PEF - Propane Equivalency Factor

PEM - Proton Exchange Membrane

PFI - Port Fuel Injection

PGM-FI - Programmed Gas Management Fuel Injection

PID - Parameter Identification Location

PKE - Passive Keyless Entry

PNP - Positive Negative Positive

POT - Potentiometer

PPE - Personal Protective Equipment

PPM - Parts Per Million

PPS - Accelerator Pedal Position Sensor

PROM - Programmable Read-Only Memory

PSI - Pounds Per Square Inch

PTC - Positive Temperature Coefficient Resistor

PTO - Power Take Off (4WD Option)

PUWER - Provision and Use of Work Equipment Regulations

PWM - Pulse Width Modulation

RAM - Random Access Memory

RBS - Regenerative Braking system

RCM - Reserve Capacity Minutes

RDS - Radio Data System

REF - Reference

RFI - Radio Frequency Interference

RHD - Right Hand Drive

RIDDOR- Reporting of Injuries Diseases and Dangerous Occurrence Regulations

RKE - Remote Keyless Entry

RMS - Recovery Management Station

ROM - Read Only Memory

RON - Research Octane Number

RTV - Room Temperature Vulcanizing

RWD - Rear Wheel Drive

SAE –Society of Automotive Engineers (Viscosity Grade)

SAI- Swivel Axis Inclination

SCR- Selective Catalytic Regeneration

SCS - Sick Car Syndrome

SFI - Sequential Fuel Injection

SI- Spark Ignition

SIPS - Side Impact Protections System

SOC- State of Charge

SOHC - Single Overhead Cam

SPFI - Single Point Fuel Injection (throttle body)

SRI - Service Reminder Indicator

SRS - Supplementary Restraint System (air bag)

SRT - System Readiness Test

STFT - Short-Term Fuel Trim

SWB - Short Wheel Base

SWL- Safe Working Load

TAC - Throttle Actuator Control

TACH - Tachometer

TBI - Throttle Body Injection

TC - Turbocharger

TCC - Torque Converter Clutch

TCM - Transmission Control Module

TCS - Traction Control System

TD - Turbo Diesel

TDC - Top Dead Centre

TDI - Turbo Direct Injection

TOOT- Toe Out On Turns

TP - Throttle Position

TPM - Tyre Pressure Monitor

TPP - Throttle Position Potentiometer

TPS - Throttle Position Sensor

TSB - Technical Service Bulletin

TV - Throttle Valve

TXV- Thermal Expansion Valve

UART - Universal Asynchronous Receiver-Transmitter

UJ- Universal Joint

USB - Universal Serial Bus

UV - Ultraviolet

V - Volts

VAC - Vacuum

VAF - Vane Airflow Meter

VDP- Variable Diameter Pulley

VDU- Visual Display Unit

VIN - Vehicle Identification Number

VPE- Vehicle Protection Equipment

VSS - Vehicle Speed Sensor

W/B - Wheelbase

WOT - Wide Open Throttle

WSS - Wheel Speed Sensor

YRS - Yaw Rate Sensor.

INDEX

Absorbed Glass Matt
   AGM 88
Actuator 68
Actuators 132
aerodynamic 121
aesthetics 121
air-conditioning 124
All electric vehicles
   AEV 77
alternating current (AC) 22
Ammonia Green (NH_3) 169
Amp 30
amp hours
   Ah 30, 87
amplitude 52
amps clamp 48
amps draw 62
anaerobic digester 163
Analogue multimeters 43
anthropogenic 132
Arc blast 73
Arc flash 73
armature 94
Atkinson cycle 118
atomised 150
atoms 23
attract 27
autoranging multimeter 45
auxiliary 41
AVO meter 43
Bad earth 60
battery 85
Battery control module
   BCM 108
battery electric vehicles
   BEV 77
Bioalcohol/Ethanol 168
Biodiesel 164
Biogas 163
biogenic 163
Brake by wire 84
Brush type AC motors 96
brushes 95
Burns 73
bus bar 107
butanol 169
cadmium 89
calorific value 135
capacitors 93
Carbon dioxide – R744 124

Carbon dioxide (CO_2) 138
carbon footprint 132
Carbon monoxide (CO) 138
CE mark 7
cells 85
cetane 168
chafed 61
charged 28
circuit 26
Circuit Breaker 75
Collaborative 82
Collaborative Motor Drive 82
Combination hybrid 81
commutator 95
Compressed natural gas (CNG) 162
Compression 117, 145, 148, 152, 153, 156
compression ignition (CI) 143
compressor 153
concave 123
conductor 25
connectors 103
consumers 34
continuity 26, 50
continuously variable transmission
   CVT 121, 122, 123
Control of Substances Hazardous to Health
   COSHH 9
controlled waste 14
Controller Area Network
   CAN-Bus 68
Crankcase ventilation 177
Cranking amps 88
current 22
cylinder bore 144
DC to DC converter 109
dead short 61
diagnostic trouble code
   DTC 54
Dichlorodifluoromethane – R12 124
Digital multimeters 43, 44
Digital principles 68, 132
diode 50
direct current (DC) 22
drag 121
Duty cycle 68, 132
earth return 103
ecotoxic 14
ECU 67, 132
efficient 171
Electric shock 73

electrical continuity 103
Electrical units 29
electricity at work regulations 1989 10
electrolysis 174
electrolyte 86
electromagnetic interference
   EMI 48
electromotive force 30
electrons 23, 179
EMF 30
Employers' Liability (Compulsory Insurance) Act 1969 11
engine braking 120
Environmental protection 14
Environmental Protection (Duty of Care) Regulations 1991 15
Environmental Protection Act 1990
   EPA 14
E-OBD 55
epicycle 122
Ethanol 168
Exhaust 117, 128, 136, 137, 138, 139, 146, 148, 152, 154, 157
Exhaust emission standards 139
Exhaust emissions 136
external combustion engines 143
field coils 95
fire extinguishers 16, 17
Fire safety 16
first aid 11, 18, 20, 21
flash point 167
fossil fuels 133
four-stroke operating cycle 144
frequency 51, 53
fuel cell 178
full-scale deflection
   FSD 43
generator . 27, 66, 80, 82, 83, 87, 91, 100, 105, 109, 163, 178
Global warming
   GWP 141
hazardous waste 16
hazards 12
Health and Safety (Display Screen Equipment) Regulations 1992 11
Health and Safety (First Aid) Regulations 1981 . 11
Health and Safety Executive
   HSE 6
heart fibrillation 74
hertz 51
High resistance 59
High voltage cabling 105
Hot cranking amps
   HCA 88
Hybrid drive 79

hydrocarbons
   HC 133, 138, 160
Hydrogen 174
Hydrogen internal combustion engines (HICE) 175
Induction 96, 98, 117, 144, 148, 151, 153, 156
Induction type AC motors 98
insulated gate bipolar transistors
   IGBT 110
insulator 24
Integrated Motor System 82
inverter 99, 110
invertion 68
isolation 114
Lambda window 137
lead acid 85
lead peroxide 85
legislation 6
Liquefied petroleum gas (LPG) 56, 158
liquid crystal display
   LCD 43
Lithium Ion 92
low carbon technologies 132
Low pressure direct injection two-strokes 150
Magnets 27
manifold 146
Manual Handling Operations Regulations 1992 11
Methanol 168
molecules 174
Motor control module
   MCM 109
Multimeters 42
multiplex 68
natural gas vehicles (NVG) 162
Networking 68
neutrons 23
Nickel cadmium 89
nickel hydrate 89
Nickel-Metal Hydride
   Ni-MH 89
Noise at Work Regulations 1989 11
octane 168
Ohm 30
open circuit 58, 104
organic matter 163
original equipment manufacturer
   OEM 65, 131
oscilloscope 52
Oxides of nitrogen (NOx) 138
Parallel 35, 78, 80, 181
Parallel circuits 35
Parallel hybrid 80
Parasitic drain 62
Pd 30
Peak oil 133, 136

periodic table 24
Personal Protective Equipment
    PPE 4
Personal Protective Equipment (PPE) at Work Regulations 1992 7
phase 99, 155
plug-in 77
polarity 40
polyalkylene glycol
    PAG..126
ports 147
potassium hydroxide.....................................89
potential difference 30
Power 3, 39, 109, 117, 146, 148, 152, 153, 157
Power control unit 109
Power probe 39
Pre-compression 148
pre-ignition 175
prohibition notices 6
propanol...169
propulsion 171
proton exchange **membrane** (PEM) 179
protons 23
Provision and Use of Work Equipment Regulations 1998
    PUWER...8
pulse width modulation
    PWM 68, 132
quenching..138
R134a 124
reasonably practicable................................12
recovery position 19
rectification...68
regenerative 83, 84, 100, 109, 120
repel 27
Reserve capacity minutes
    RCM..88
Residual Current Devices.............................75
resistance 30
retrofitting..158
risk assessment 13
risks 12
rotary engine...155
rotor 96
scan tool 54
scavenging 148
semiconductor 172
Sensors..67, 132
Series. 34, 78, 80, 181
Series circuits 34
Series hybrid 80
Short circuit..60, 61

solar 172
spark ignition (SI) 143
starting 100
state of charge
    SOC..91
static electricity 22
stator 96
steam engines 143
stoichiometric 137
supercharger 153
Synchronous type AC motors......................96
terminals 103
Test lamps 37
Tetrafluoropropane - R1234 yf....................124
The Health and Safety at Work Act 1974
    HASAWA..7
The Health and Safety Information for Employees Regulations 1989.................11
The
    Management of Health and Safety at Work Regulations 1999 11
The Pressure Safety Systems Regulations 2000 11
thermal energy 135
Thermistors..91
Three phase type AC motors.......................99
Toroidal CVT 123
torque..157
torus 123
traction motor 78
Transesterification.....................................165
Transfer...148
transfer note 15, 16, 113
transistor 51
Transmission 121
trickle charge 91
two-stroke operating cycle 147
type approved 56
vaporise 145
Variable diameter pulley
    VDP...122
vehicle protection equipment
    VPE..5
ventricular fibrillation 74
viscosity..165
Volt..30
VTEC 119
Waste vegetable oil 165
Watt...31
wiring...101
Workplace (Health, Safety & Welfare) Regulations 1992 11

www.ingramcontent.com/pod-product-compliance
Ingram Content Group UK Ltd.
Pitfield, Milton Keynes, MK11 3LW, UK
UKHW051351180426
11947UKWH00014B/879